EFFECTING A QUALITY CHANGE:
AN ENGINEERING APPROACH

EFFECTING A QUALITY CHANGE: AN ENGINEERING APPROACH

S W Field

Consultant in Quality Management
formerly Quality Director
British Aerospace Military Aircraft Division

K G Swift

Lucas Professor of Manufacturing Systems Engineering
Department of Engineering Design and Manufacture
University of Hull

A member of the Hodder Headline Group
LONDON • SYDNEY • AUCKLAND
Copublished in North, Central and South America by
John Wiley & Sons, Inc., New York • Toronto

First published in Great Britain in 1996 by
Arnold, a member of the Hodder Headline Group,
338 Euston Road, London NW1 3BH

Copublished in North, Central and South America by
John Wiley & Sons, Inc., 605 Third Avenue,
New York, NY 10158-0012

ISBN 0 470 236884 (Wiley)

Whilst the advice and information in this book is believed to be true and
accurate at the date of going to press, neither the author[s] nor the publisher
can accept any legal responsibility or liability for any errors or omissions
that may be made.

British Library Cataloguing in Publication Data
A catalogue record for this book is available from the British Library

Library of Congress Cataloging-in-Publication Data
A catalog record for this book is available from the Library of Congress

ISBN 0 340 676515

Typeset in 10/12pt Times by Wearset, Boldon, Tyne and Wear.
Printed and bound in Great Britain by J.W. Arrowsmith Ltd, Bristol.

Contents

Preface

Manufacturing businesses worldwide are facing fierce competition and operating in changing markets. Customers are demanding higher quality and ever more competitive prices and delivery times. Countries that were once profitable stable markets for products are now competitors. We need only recall the UK's former world position in the production of motor cycles, textiles, shipping and consumer electronics, to indicate that industry cannot relax but must continuously strive to improve its competitive edge.

Why does a company need to produce a profit? Examine any company's annual report and you will see the need for profit for:

- shareholder dividend;

- growth and position in the market;

- development of new products and processes;

- workforce benefits, etc.

We intend to investigate and determine what is quality, what we mean by quality and how it crucially affects the profit and competitiveness of a company.

The benefits to a business for improving quality can be enormous. Given that it is not so uncommon for a business to have a total cost of quality failure as high as 10% of total revenues, a company with a £2.5 billion turnover might well be losing £250 million per annum. Even a small percentage improvement is worth millions.

The text is organised in the following way. *Chapter 1* covers the modern concept of quality. The history of quality, the quality gurus and the Japanese influence, the modern concept of quality compared to the traditional approach will be investigated and a working definition of the term 'quality' established.

Chapter 2 explores quality culture; involvement of all the workforce led by the management, the need for education and training, the concept of 'teaming' of people by concurrent engineering or simultaneous engineering/natural work groups/quality circles will be reviewed, and the need for the establishment of a clearly defined policy and organisation.

Chapter 3 is concerned with total quality. The evaluation of total quality (to improve efficiency and maximise profit by a total company involvement) and the concept of both internal and external customer relationships is reviewed. The need for international/national standards (e.g. BS EN ISO 9000 – 'Quality Management and Quality Assurance Standard') are discussed.

Chapter 4 addresses external supplier quality. In most manufacturing organisations more than 60% of the final product is 'bought in', hence there is a need to develop a supplier strategy. This strategy must be rigorously controlled, but should not take the responsibility away from the supplier, from specification through to design, qualification, manufacture, testing, delivery and maintenance of standards.

Chapter 5 concentrates on software quality control. Many companies rely on software for the operation of their business e.g. computer-aided design (CAD), financial control, factory management, numerical control (NC) manufacture, etc. Equally, many high-technology products, e.g. motor vehicles, aerospace and electronics, rely on software for their operation. It is vital that the software be error free and configuration controlled.

Chapter 6 introduces quality tools and techniques. A company must be prepared to give its workforce continuous quality education and training if it wishes to improve its quality. This is not a one-shot operation but a continuous process. There are a number of tools and techniques such as Pareto analysis, cause-and-effect diagrams, workflow analysis, etc., which are an essential part of the training and business practice. Also presented is a general framework showing where in the product introduction process the various tools and techniques should be applied.

Designing for quality is covered in *Chapter 7*. A major objective of the book is to illustrate the importance of 'Design for Quality' (DFQ). Quality cannot be inspected into a product, it is designed into the product. The major elements include an adequate specification and specification control, adequate R&D and project definition, supplier quality, design reviews, configuration control and traceability and quality-improved planning.

Chapter 8 reviews the costs of quality. Surprisingly very few companies have ever attempted to determine the true cost of quality. This may be due to a number of factors, e.g. unwillingness to indicate failures or the fact that many cost-control systems do not measure the cost of failures, which should include: late issue of design, tool errors, out-of-sequence working, excess work in progress, shortages on assembly, excess inventory, poor supplier quality, penalty clauses for late delivery, etc. A series of costing systems are outlined and the concept of value- and non-value-added operations discussed.

Chapter 9 is a review of selected quality tools and techniques. Some of the main quality tools and techniques referred to in Chapter 6 are described more comprehensively in this section, together with case studies.

How to effect a quality change is the topic of *Chapter 10*. How does a company embark on a program to improve its quality? It is vital that the initiative starts with the chairman/managing director. If he/she is not leading the crusade, it will fail. A suggested procedure comprises a review of the company quality by individual departments, followed by the presentation of a continuous education, training and planning program. It is imperative that metrics of performance are developed, objectives are set and progress is reviewed. A change cannot be introduced overnight. In large companies it is possible that it may take as long as five years to be fully effective.

Chapter 11 touches on future developments. Significant changes are underway in Europe, USA and Japan which must be carefully monitored. Any future developments to improve quality should include a co-ordinated research programme to develop further DFQ techniques, growth in Total Quality Management (TQM) and improved software quality.

The book will be helpful to university staff and practicing engineers and managers interested in effecting a quality improvement. It also provides undergraduates and postgraduates of engineering and business studies with course and project material in quality management and quality engineering.

The authors are greatly indebted to Julian Booker, Researcher in product quality engineering, of the Department of Engineering Design and Manufacture, University of Hull, for his invaluable help with the preparation of the book, particularly in the areas of designing for quality and quality tools and techniques.

British Aerospace Military Aircraft Division is gratefully acknowledged for its permission to use material from its Continuous Quality Improvement (CQI) training manual included in the text. The following individuals have been especially helpful: Alan Millican, Quality Director, and John Whalley, Quality Assurance Director.

We are grateful to Joe Cullen, Operations Strategy Director and Alan Curtis, Managing Director of Product Supply at Rover Group for permission to use material relating to total quality.

Thanks are also due to Clive Kinder and Jim Tattersall of I.C.O.M. for their work in the field of quality training.

Richard Batchelor, Manufacturing Design Manager of Lucas Industries (Electronics Division), is thanked for his invaluable contributions and enthusiastic support of the work on design for quality.

Thanks are also due to Bob Swain, of the University of Hull, for his help with the preparation of this manuscript.

The work reported here on design for quality is supported by the Engineering and Physical Sciences Research Council under research grant GR/J97922.

The approach put forward in this book for effecting a quality change in a business is applicable to large and small companies alike. While most of the quoted industrial experiences are based on large organisations, this does not imply that they are only applicable to large companies. The common ingredient is the people and their commitment to a cultural change.

1 The modern concept of quality

1.1 Historical perspective

In the UK prior to the Industrial Revolution, manufacture was essentially conducted by the 'cottage industry' approach and relied heavily on craftsmen. The craftsmen were organised into Guilds which are still evident in some of the City Guilds today. The craftsmen trained apprentices to ensure the quality and standard of work (Figure 1.1). However, with the advent of the Industrial Revolution the skilled men could not possibly cope with the tremendous increase in manufacture required.

The problem was solved by breaking the work into smaller jobs permitting a large, mainly unskilled workforce to perform limited skill operations

- Craftsmen organised into Guilds

- Craftsmen trained their apprentices to maintain and control skill levels

- Craftsmen did everything from raw materials procurement to after sales service

Multi-Skilled, Self-Certified Workforce

Fig. 1.1 Pre-Industrial Revolution

- Manufacture broken into small jobs

- Introduction of a large unskilled workforce

- Use of craftsmen as "overlookers" (inspectors)

- Emergence of Standards (e.g. Whitworth threads)

Fig. 1.2 Industrial Revolution

(Figure 1.2). The craftsmen became overlookers (inspectors) and the first standards were introduced, e.g. screw threads, hole sizes, etc. In a similar manner the two world wars demanded a further rapid expansion in manufacturing potential. This was again accomplished by the use of a largely unskilled workforce, standards and the introduction of the inspector system with perhaps one inspector per ten operators in some industries. The problem with many industries is that they have not progressed from that position and still rely on the use of inspectors to ensure the quality of the product leaving the factory (Figure 1.3).

This highly inefficient approach of detection and either repair or rejection is unaffordable. Quality is the responsibility of everyone, from the

- War meant a rapid expansion of the workforce

- Use of unskilled workforce strengthened the need for "Inspection" after every operation. Skilled workers as Inspectors

  Todays problem:

- We are still living with this legacy of manufacturing

Fig. 1.3 Effect of two world wars on manufacturing in the West

R&D stage through project definition, design, manufacture and customer support.

The initial ideas of a new approach to quality emerged in the USA during World War II with a statistical approach to assist quality control. However, after the war there was a universal demand for goods and services. Markets were buoyant with little competition. An age of complacency followed with emphasis on output with insufficient regard for quality, particularly in design and customer satisfaction.

Two of the quality pioneers or gurus, W. Edwards Deming and Joe Juran, with their statistical and process approval and systematic approach to quality improvement projects, are acknowledged for the part they played in the post-war Japanese quality revolution (Figure 1.4). The Deming Prize, first presented in 1951, is still considered to be the highest award relating to statistical quality control in Japan today.

In the 1960s the need for improvement in quality was recognised in the UK, particularly by the British Standards Institution (BSI) and the Defence Industry. There was general support for the role of Quality Assurance, e.g. prevention rather than correction, quality planning and training, development of repetitive processes to ensure a common standard, improved tools, etc. (Figure 1.5).

1940–45
- Americans pioneered the use of statistics to assist the Inspection process
- Deming and Juran were two of the experts

1945
- Mass markets of the West waiting to be filled
- Skillful quality management took a back seat

1950's
- Buoyant markets and little competition
- Output targets the only goal
- Quality and productivity ignored
- Age of complacency
- Deming and Juran became yesterday's men
- Deming went to Japan, his ideas were adopted with passion
- Japanese started the move away from specialist quality staff
- Introduced the concept of quality in non production areas
- 25 years later they overtook the West

Fig. 1.4 Post-World War II

1960's

- Advent of Quality Assurance in British industry
- Concept of QA restricted to preparation of procedures and training and assurance of proper tools, materials and equipment to the shop floor

1980's

- Quality is still regarded by some as solely the responsibility of the quality department
- Introduction of quality standards (BS EN ISO 9000)

1990's

- Greater awareness of the importance of quality to achieve competitive performance
- Understanding of the need for presentation and interest in quality in product engineering

Fig. 1.5 Quality in British manufacturing

Since that time many companies have made enormous improvements but there remains much to be done (Figure 1.6). The graph illustrates how

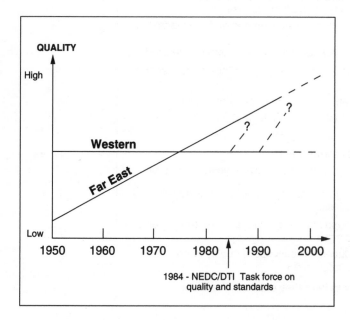

Fig. 1.6 World competition in quality

the Japanese, starting from a very poor base, overtook the West in the mid-seventies and continued making steady progress. On the other hand, many Western companies made little progress and required significant improvement to catch Japanese industry. A 1988 White Paper on employment in the 1990s stated, 'In the 1990s what will distinguish successful businesses and countries from the rest is how effectively they use the skills and abilities of their people to meet the needs of the market and how quickly they respond to new pressures'.

Since 1988, consultation grants have been given to small and medium-sized companies by the Department of Trade and Industry as part of the Enterprise Initiative; more than half of the applications relate to quality assurance. The British Quality Association makes an annual British Quality Award and supports the spread of quality-assurance systems.

1.2 International view

The European Foundation for Quality Management (EFQM), the European equivalent of the British Quality Foundation (BQF), was initiated in 1988 and by 1995 over 350 companies had become members. In 1992 EFQM jointly launched the European Quality Award with the European Commission and the European Organisation for Quality. The award is given to companies who have been successful exponents of Total Quality Management (TQM) and who demonstrate excellence in the management of quality as their fundamental process of continuous improvement.

EFQM enables a company to perform self-assessment against benchmarks across a number of enabling parameters:

- leadership;
- people management;
- policy and strategy;
- resources;
- processes;

and results parameters:

- staff satisfaction;
- customer satisfaction;
- impact on society;
- business results.

In Germany, some quality-assurance issues are already being dealt with on a project-related basis in some of the technical programmes being run by the Federal Ministry for Research and Technology, e.g. the manufacturing and space programmes.[1]

In the context of industrial co-operative research, the German Federal Minister for Economic Affairs provides selective support for quality-assurance-related projects via the Association of Industrial Research Organisations. This support began in 1989 with a quality-assurance scheme. Funding is provided for research projects at research institutions which are sponsored by several industries or sectors; the projects must be geared to small- and medium-sized businesses and offer the possibility of implementation in a wide variety of circumstances. They cover subjects such as the analysis of specific production processes with the aim of developing reliable, robust techniques by clarifying the connections between process parameters and product quality.

Other areas of study include the development of measuring, testing and evaluation techniques as elements within quality-assurance systems and quality-assurance strategies, or the question of the economic benefit of preventive quality-assurance systems in certain industries or in the processing of certain materials. DM22 million has been approved for 50 projects since 1989.

Several institutions are active in the field of training and continuing education, e.g. the German Quality Association and the German Industry Rationalisation Board, which are successfully involved in specialist training and publish a wide range of quality-related literature.

It is also worth mentioning the French Movement (Francais pour la Qualité) which since March 1991 has brought various state and private agencies and organisations together in a multidisciplined group. Its measures to promote quality are primarily intended to prepare small- and medium-sized businesses to meet the requirements of the European internal market.

The nature of the initiative taken by the US Government to highlight the importance of quality is interesting. In August 1987, Congress passed the Malcolm Baldrige National Quality Improvement Act of 1987 in which the importance of quality for the competitiveness of American industry is emphasised and annual awards offered.

Other public measures to promote quality improvements are included in the programmes of various agencies of the US Federal Administration. These were brought together in an interdepartmental initiative on production technology in 1993.

Something of the US renaissance[2] and more on the historical and international perspectives can be found elsewhere.[3]

1.3   What is quality?

The achievement of quality must also be linked with cost. It is no use achieving the required quality at a cost that will not be competitive and it is equally no use achieving competitive cost by degrading the quality, thereby eroding customer satisfaction.

A successful company must continually satisfy the following three demands:

- delivery on time;

- cost as budget;

- quality as required by the customer.

This is represented diagrammatically in Figure 1.7.

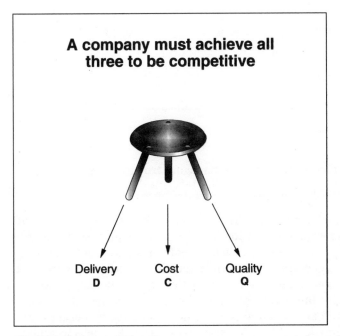

A company must achieve all three to be competitive

Delivery   Cost   Quality
    D   C   Q

Fig. 1.7 Delivery, cost and quality

We must therefore concentrate on prevention not inspection, introduce quality in design, control the process and give responsibility to staff (Figure 1.8).

The quality systems must be geared to prevention rather than detection and correction. The concept of detection is wasteful. It is not foolproof and does not add any value to the product or service.

Prevention is better than detection because:

- detection is a waste – it costs money and doesn't add value;

- detection is not foolproof;

- detection de-motivates.

Prevention means:

- tackling the cause, not the effect;

Traditional way:

- Tackle quality problems by increased inspection

New approach:

- Quality in design
- Control in process
- Give responsibility to staff
- Right first time

Fig. 1.8 Prevention not inspection

- solving the problems at source, not managing around them;

- removing the problem for good, not just this once.

To bring about a change in an organisation, it is vital that all the staff clearly understand what a prevention-based system entails and that change is led from the top. To be fully effective a quality system must be designed to prevent defects occurring rather than reacting to them after the fact. That is to say a prevention system is better than a detection system. Continuous Quality Improvement (CQI) focuses on the fact that quality cannot be inspected and repaired into a product. It must be designed and built in, in a systematic and fully integrated way. The attitude is to achieve 'right first time'. Quality can be defined as:

- fitness for the intended purpose;

- meeting the agreed specification;

- ensuring that the product or service is fully acceptable to the customer;

- customer satisfaction.

The achievement of quality is not a one-off act but a continuous process. In short, quality is defined as fit for intended purpose and conformance to agreed customer requirements. It is not an absolute measure, but a constantly moving target, generally perceived to have two major aspects:

- the degree to which goods and services meet customer needs and therefore lead to repeat business;

- the extent to which the conversion process is free from product and operational deficiencies which should lead to making appropriate margins.

1.4 Quality principles

Out of the work of some of the well-known quality gurus, namely Deming,[4] Juran,[5] Ishikawa[6] and Crosby,[7] 13 principles have evolved. These have been applied with varying degrees of success. The principles are listed below:

- innovate in all areas including training and provide resources to assist. Maintain an innovative and vigorous training programme;
- learn the 'zero-defect' philosophy and the need for continued improvement;
- do not rely on mass inspection for quality. Put quality prevention on line via statistical process control;
- reduce the number of suppliers and develop them for continuous improvement of service as well as cost;
- use statistical techniques to identify sources of waste and cure both system faults and local faults at source;
- ensure that organisational and management systems support innovation and continuous improvement;
- provide supervision with on-line techniques for problem identification and problem solution via their teams;
- create openness by encouraging questions and then reporting the problems;
- attack waste by the use of multidisciplinary teams;
- avoid exhortative slogans as a substitute for team approaches;
- beware of over-bureaucratic imposition of work standards;
- provide elemental statistical training for all employees;
- make maximum use of statistical data to focus on priority problems and direct the efforts of all the talent in the company.

The field of quality management, perhaps more than any other, has been strongly influenced by a small number of gurus. Their opinions do not always agree and a business needs to examine where they are on the performance scale and take on board the views that are most appropriate to its situation. To provide a more complete picture, the reader is directed to the works of Feigenbaum,[8] Shingo[9] and Taguchi.[10] An overview of the messages of the quality gurus can be found in Bendell's works.[11]

Before leaving the topic of quality gurus, it is worth noting that whilst each may be associated with a particular message, i.e. Crosby with 'zero defects' or Deming on the 'responsibility of management', the authors believe the unifying philosophy to be the people and operating cultures.

2 Quality cultures

For many years poor quality was generally associated with shop-floor operators. It is not generally appreciated by many businesses that most of their quality problems are really management problems. Over 75% of company quality deficiencies can be traced to management-related deficiencies (not shop-floor activities). Figure 2.1 is a collection of comments from quality leading international businesses who tend to support this view.[12]

"...over 75% of product engineering rework is due to management problems..."

"...85% of all faults detected in products originate during product development and planning..."

".. 80% of faults are only detected in finished parts or products and when they are in use by the customer..."

"...ISO 9000 (alone) will not increase our abilities to compete..."

"...To be competitive we must think quality first. When quality is right everything else will fall into place including cost and productivity..."

"...We need a culture, professionalism and tools for Design for Quality (DFQ)..."

Fig. 2.1 Comments from quality leaders

Modern quality demands:

- everyone responsible for quality;
- management leadership;
- challenge to existing management practices;
- quality not cost as the first priority;
- right first time;
- customer satisfaction.

A focus on quality is shown in Figure 2.2.

Over the years many companies, particularly large organisations, have drifted toward 'four-walled management'. The companies develop powerful, largely autonomous, functional departments, e.g. marketing, technical, commercial, manufacturing, finance, personnel, customer support, etc. The problems associated with these organisations are:

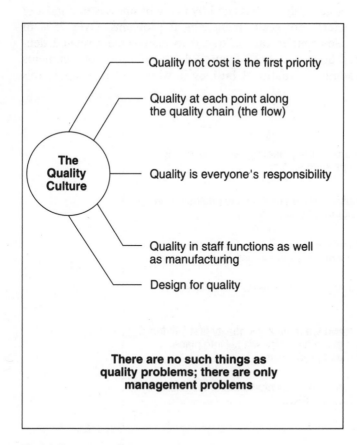

Fig. 2.2 Focus on quality

- departments over-specialised and fragmented;
- competition rather than co-operation;
- non-value-added activities and systems;
- scapegoats;
- errors not pinpointed;
- slow to change;
- slow to solve cross-functional problems;
- excessive co-ordination and paper work;
- poor lateral communication;
- multi-skilling discouraged.

2.1 The need for teamwork

During the last decade many companies have introduced, with varying degrees of success, ways of improving operations by the concept of the natural grouping of people, for example:

- concurrent engineering;
- natural work teams;
- quality circles;
- quality action teams, etc.

These alternative groupings feature:

- division of added-value chain into 'natural works groups' – a relatively independent and significant set of activities that naturally cluster together;
- de-centralisation of support functions;
- creation of multi-skilled cells;
- creation of business units.

Quality circles[13-16] were introduced to the UK in the 1970s; they were the 'in thing', the latest management task. However, in many cases they were unsuccessful compared with their success in Japan, the USA and India.

What is a quality circle? A useful working definition is as follows[16]:

Quality circle is a small group of employees in the same work area or doing similar type of work who voluntarily meet regularly for about an hour every week to identify, analyse and resolve work-related problems leading to improvement in their total performance and enrichment of their worklife.

2.1.1   Why did quality circles not succeed?

Any quality improvement campaign or initiative must start from the top of an organisation and cascade down, starting with policies/strategies. In many cases in the UK, quality circles started on the shop floor, and consequently the middle management were deeply suspicious and in some cases viewed the teams as a threat to their authority. Any identified improvements could be regarded as a failing of the middle management. Do not underestimate the importance of gaining the commitment of the middle management!

When teaming staff on quality circles, it is important to:

- ensure that initiatives start from the top of the company;

- ensure the involvement and commitment of all management;

- ensure the commitment of first-line supervisors;

- train circles and use facilitators;

- brief staff and unions;

- recognise that quality circles are only one means of improving quality;

- be prepared to give any skills/numeracy training the circles identify;

- start with pilot operations;

- monitor results.

Quality Action Teams (QATs) and Quality Improvement Teams (QITs) are now operating in many UK companies. They usually adopt the quality circle principles except that any quality improvement projects are led by management deciding where and when the introductions should be made, what resources in terms of staff and capital can be afforded, agree time scales for implementation and audit the benefits. QATs are discussed in more detail later.

The advantages of the natural grouping of people are:

- total ownership and responsibility for the whole project by a team;

- removal of a high percentage of non-value-added or waste activity;

- efficient communication – people talk across a table;

- simplicity of flow for planning purposes – planning the overall team requirement, not the work of each specialist;

- the team can affect and have control over priorities, effectiveness and responsiveness;

- shorter lead times and high reliability of the process carried out by the team;

- multi-skill role development; aids job satisfaction and flexibility, whilst enhancing career opportunities;

- attracts reduced overheads and supporting infrastructure;

- facilitates a clear definition of the role; supports needs, inputs, outputs and measures of performance;

- clear allocation of responsibility for the quality of team performance;

- makes definition of customer/supplier relationships between individual teams easier to design and operate;

- allows effective use of SPC and FMEA (see Chapter 9) on the team process – note that FMEA is also applicable to office processes.

2.2 The role of the quality department

Many companies are now developing more 'lean' and efficient organisations based on very positive objectives, e.g. no-one in a management position should have less than six, or more than ten people reporting to them; everyone should have a clear job description and there should be no more than five layers of management from top to bottom.

These are good principles; however, many companies find difficulty in understanding the role of the quality department. In many cases the quality department is erroneously regarded as just another department. Ideally the quality department should be independent of all other departments and directly responsible to the managing director. Each functional department must be responsible for its own quality, usually with the help of a department quality controller responsible to the departmental head. The quality department is then responsible for policy, strategies, guidance, consultancy, help, training and planning for the whole company. The quality director or executive is ultimately responsible for company quality next to the managing director and must have the last word on any decisions (Figure 2.3).

Now let us define what we mean by culture. Culture is the set of attitudes that people at work share with each other and which affect how they do things on a day-to-day basis. They transfer these attitudes to new recruits via training practices. Culture affects behaviour, behaviour affects quality and quality affects business success.[17-19]

Culture has a direct effect on total quality (see Chapter 3):

- Attitude   'Quality is someone else's responsibility.'
  Results   People let mistakes go. They don't check work.

- Attitude   'No one listens to what we say.'
  Results   Low involvement in quality improvement schemes.

- Attitude   'Management are there to help me achieve my goals.'
  Results   Active support for total quality initiatives.

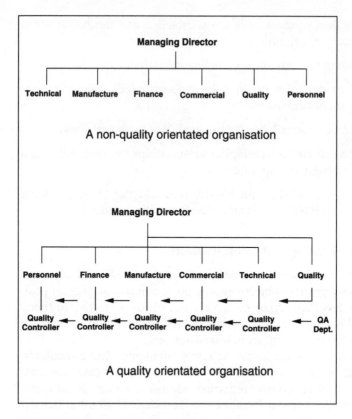

Fig. 2.3 Contrasting organisation structures

A quality company ensures that:

- profits come from customer satisfaction;
- problems are prevented rather than detected;
- complaints are used as opportunities to learn and improve;
- people and teams are recognised as important elements in product and process development;
- training and personal development are a priority;
- every activity should contribute to the quality of performance;
- everybody contributes to quality;
- quality service is provided for all customers – internal and external.

A quality company exhibits the following cultural influences:

- company mission;
- the company's technology;

- organisational structures;

- organisation systems;

- management philosophy and style;

- education and training.

In the following industrial example we will touch on the policy and quality goals set by the management of British Aerospace Military Aircraft Division (BAe MAD). (The notes are based on an article produced by Alan Millican, Quality Director of BAe MAD.[20])

2.3 Industrial experiences – our commitment to quality

To quote:

> The last major review of quality within the Division was back in 1988. Since that time our approach to Quality has developed and matured in line with best practice and the changing business environment of our industry. What has not changed is the understanding that the application of Quality principles has the potential to make a substantial contribution towards better business performance.

Against this background it is essential that we all have a common understanding of the quality dimension within MAD and how we can use the tools of quality to achieve competitive advantage. Consequently, over the past 12 months, we have been reviewing our quality policy and strategy to ensure that it lines up with our current thinking and will remain relevant as we continue to reshape the business to meet the challenges of the late 1990s.

The outcome of the review has become twofold:

- a restatement of our quality commitment and the positioning of 'customer satisfaction at competitive cost' as our prime quality goal;

- a new quality policy which brings together our current understanding of quality into a series of policy statements.

The new policy focuses on six key elements:

Business excellence: The use of the European model for total quality management brings a sharper definition to the concept of total quality as it is clearly focused on delivering business excellence.

Quality assurance: The commitment to the ISO 9001 standard has been linked to quality planning and prevention as the prime focus for our regulatory and internal quality-assurance activities.

Partnerships: The importance of securing partnerships with both our customers and suppliers in the drive for improved quality is recognised. Our preferred supplier process and the drive to secure MacDonald Douglas Aircraft approved supplier status are examples of our commitment to partnerships.

Processes: The policy acknowledges the importance of understanding, controlling and improving our processes. Business Process Re-engineering (BPR) and the application of Statistical Process Control (SPC) are our current major activities.

People: The Continuous Quality Improvement (CQI) programme, introduced in 1990, substantially increased our awareness of quality and associated training programmes gave people additional skills that helped them make a positive contribution in performance improvement initiatives. Building on this base, quality-led activities are now more sharply focused to give a clear business benefit, but at their heart is still the CQI principle of involvement. Quality isn't the domain of the quality director alone, it belongs to everyone. The policy contains a statement on quality ownership and a clear commitment to ensure everyone is fully equipped to play their part in securing our quality goals.

Continuous improvement: This is the underpinning concept for the quality policy. It implies learning to lock-in experience gained, and with our complex business a structured approach is required; hence the need for processes that embody the concept of target, measurement and review. At the top level the business plan embodies this concept, at the individual level the suggestion scheme rewards ideas designed to deliver measurable benefit. Of course, in line with our policy of continuous improvement, our approach to quality will evolve and mature, but we now have a framework which should enable us to build upon the progress we have made over recent years.

To support the Division's drive for improved performance we have set ourselves the following strategic quality goals:

- achieve 'world class' standard for quality, cost, service and value for money using the European Foundation for Quality Management (EFQM) model as a bench-mark;

- reduce the cost of quality to less than 10% of cost base;

- all basic operational processes maintained and controlled via SPC and processes regularly reviewed and continuously improved by process owners;

- the supplier base recognised as 'world class' in respect of quality;

- all critical business processes re-engineered, optimised and bench-marked as 'world class';

- quality methods and tools fully integrated into MAD's normal business processes;

- achieve 'world class' levels of employee involvement on CQI.

3 Total quality

In most companies a decision to increase their profits by 15% would be achieved by firstly increasing the sales by 15%, increasing the plant output by 15% and increasing the workforce by 15%, assuming that they are operating at maximum efficiency.

However, it is more likely that they would be able to achieve the increased profit without any increase in capital expenditure or workforce by addressing the quality of their operations, improving the quality of their designs and the quality of their suppliers. Many businesses around the world operate with a 15–20% inefficiency due to poor quality of goods and operations.

This message is now being addressed by the more enlightened companies and they are reaping the rewards.

3.1 What is total quality?

The principle of efficiently harnessing the skills of all the workforce, for example:

- marketing;
- sales;
- design;
- manufacture;
- suppliers and purchasers;
- commercial;
- personnel;

- finance;

- customer support, etc.

is referred to as *total quality*.

Total quality is the concept of reducing costs, improving overall efficiency and having a policy of customer satisfaction, right-first-time operations and continuous quality improvement throughout the company and utilising the whole workforce.

The importance of total quality is now recognised nationally with the issue of BS 7850: Part 1 (1992) and Part 2 (1994) by the British Standards Institution.

In summary, total quality can be defined as:

- continuous quality improvement;

- customer satisfaction;

- company wide – everyone involved;

- right-first-time operation;

- prevention not correction;

- quality at the lowest cost;

- management commitment and leadership.

A useful way to illustrate total quality, developed by the Rover Group, is shown in Figure 3.1, and is based on:

- leadership from the top;

- the costs of quality;

- the customer/supplier relationship;

- continuous improvement;

- involvement of everyone.

The elements of quality organisation consist of three mutually dependent items:

- the culture – total quality;

- certification (BS EN ISO 9000);

- quality management systems.

Their relationship is shown in Figure 3.2; however, an alternative view by the Rover Group is:

- the culture – total quality;

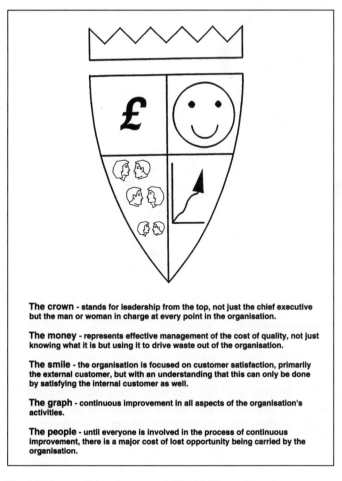

The crown - stands for leadership from the top, not just the chief executive but the man or woman in charge at every point in the organisation.

The money - represents effective management of the cost of quality, not just knowing what it is but using it to drive waste out of the organisation.

The smile - the organisation is focused on customer satisfaction, primarily the external customer, but with an understanding that this can only be done by satisfying the internal customer as well.

The graph - continuous improvement in all aspects of the organisation's activities.

The people - until everyone is involved in the process of continuous improvement, there is a major cost of lost opportunity being carried by the organisation.

Fig. 3.1 Five conditions for successful TQM (Rover Group)

- quality management systems (including certification);
- tools and techniques for quality improvement.

The achievement of BS EN ISO 9000 certification is not a guarantee of good-quality products but indicates that a company has satisfactory quality processes, systems and management in place.

Total quality involves all the organisations, all the functions, the external suppliers, the external customers and involves quality policy.

Similarly, total quality cannot be achieved without good Quality Management Systems (QMS) which bring together all functions relevant to the product or service, providing policies, procedures and documentation (Figure 3.3).

The organisations should fully understand the customer/supplier

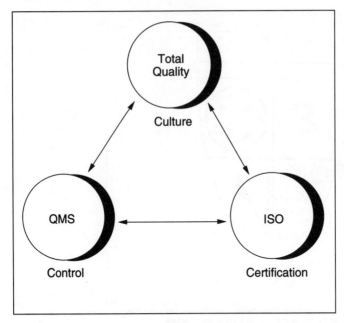

Fig. 3.2 Elements of a quality organisation

relationship. The term customer does not only refer to the ultimate purchaser of the product or service but to everyone in the company (Figure 3.4).

Everyone in the organisation has a customer to whom they pass on completed work and therefore all should be customer focused.

- Brings together all the functions, objectives and activities which contribute to the consistent Quality of the product or service

- The systematic documentation of the policies and procedures ensures the cost effectiveness of the organisation

- The documentation must be practical and up to date

- The documentation can be in manuals or computer software

- All staff must be totally conversant and ideally involved in its preparation

Fig. 3.3 Quality system

Two features of this approach need highlighting:

- The customer is not to be seen purely as the end user of the goods or services. Customers also include the person or people to whom work is passed, within the organisation, that is the internal customer

- The requirements of the customer (internal and external) should be specified in a measurable way so the extent to which they are being met is ascertainable. That is, define the code of quality.

Fig. 3.4 Customer focus

3.2 Customer/supplier relationship

Are you customer focused?

- Are you clear who your customers are?

- Do you know what they do with your outputs?

- Do you know what their precise requirements are?

- Have you asked them – tested their needs? Remember their expectations change.

- Do you get feedback from your customer?

Every person in the organisation has a supplier who passes work to them. The internal customer/supplier relationship requires those whose task performance influences product quality to produce zero-defective products for the next person/process in the internal customer/supplier chain. It also requires those whose tasks do not directly affect product quality to perform in the right-first-time mode to ensure that costs are minimised.

A general model of the customer/supplier conversion process is shown in Figure 3.5.

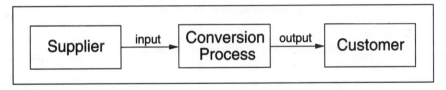

Fig. 3.5 Customer/supplier added-value chain

The objectives of total quality cannot then be achieved without knowing your position in the chain of operations. The pursuit of quality improvement means that you need to understand your role in the organisation, what your responsibilities and tasks are and how they affect your customers.

Your answers to the following questions should help you to understand the customer/supplier relationship:

- Who are you?

- What do you do?

- Who is your customer?

- What do they expect of you?

- Who supplies you?

- What do you expect of them?

- What can you do to improve?

The customer/supplier relationship can be likened to a chain (see Figure 3.6).

As can be seen from the above, total quality is broad in perspective and contains many different theories and methods. There is much published work in the field and further reading on total quality, its implementation and benefits, would be useful.[21–23]

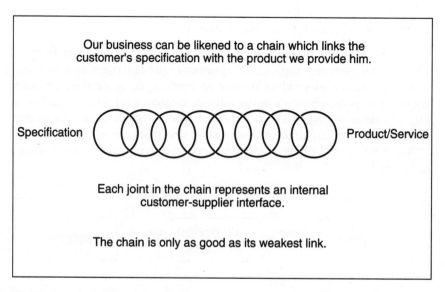

Our business can be likened to a chain which links the customer's specification with the product we provide him.

Specification   Product/Service

Each joint in the chain represents an internal customer-supplier interface.

The chain is only as good as its weakest link.

Fig. 3.6 Customer/supplier relationship

While the Baldrige Awards provide competition for US companies, they have come to be regarded as the most comprehensive statement on what total quality means and requires.[24] More information on total quality and the lessons learned from the Baldrige Awards is found elsewhere.[25]

3.3   Industrial experiences 1 – implementation of total quality at Rank Xerox

The following case study outlines the total approach to quality adopted by Rank Xerox. The initiative was called 'leadership through quality'. The review of the initiative presented here is based on a paper by Mercer and Judkins of Rank Xerox. The case is included because it still represents what is believed to be good practice if a 'total quality' approach is to be successful.

3.3.1   Changing markets

The business was facing fierce competition in all its market sectors as new competitors joined the scene. For many years the company had few rivals but more recently it was faced with a situation where there were more than 100 organisations making some 300 types of copying machine in direct competition with their own designs. Many of them were Japanese. Xerox realised that to remain an industry leader it needed to change the way it approached its work. It felt it needed to renew and enhance its commitment to the requirements of its customers. It was necessary to make the most of its people – their talents, energies and ideas. More discipline was needed in the way they managed and worked. It was recognised that the change needed could not be made overnight and a progressive step-by-step approach was required to improve its competitive position.

3.3.2   Understanding the competition

The business began the process by competitive bench-marking. This was a continuous process of measuring its products against those of its strongest competitors and renowned leaders in the field. This enabled it to identify gaps between Rank Xerox and its competitors and to establish goals in the quest to be superior in the product areas of quality, reliability and cost.

3.3.3   Unifying the work force

Having set the goals, the next step was the process of getting the Xerox employees fully involved – acting to achieve the goals. The requirement was to get the Xerox people at all levels to participate, using their talents to solve the problems identified. This was working well, but in order to enhance the process and reach the desired growth objectives the business needed to focus all the energies and talents of its people on customer satisfaction. What was missing was a focal point, the unifying force as Xerox called it, to direct the enormous energy of all Xerox employees around the world. The decision was made to move decisively towards quality as the end objective and the notion of leadership through quality was created.

3.3.4   The Xerox view of quality

Rank Xerox states that its view of quality differs from the norm in at least four major areas:

> The first is how we define it. To most people, quality conjures up words like goodness and luxury. Rank Xerox defines it as conformance to customer requirements. Quality is what the customer requires, not what we think the customer should have.
>
> The second difference is in how we set performance standards. Conventional methods of setting standards allow for some level of defect or error. For example, one defective part or installation per hundred might be acceptable. The performance standard in Rank Xerox is products and services that fully meet the requirements of our customers.
>
> The third basic difference between our view of quality and the traditional view is the system used to achieve quality. The conventional means of assuring quality is inspection, where a product is looked at after it has been completed and only quality work is passed on to the customer. By contrast, the Rank Xerox emphasis is prevention of errors.
>
> The last of the four differences between the conventional view of quality and our view is the way in which we measure it. The conventional approach to measuring quality relies on indices. Rank Xerox measures quality by the costs we incur when we do not satisfy our customer requirements (the cost of non-conformance).

3.3.5   The Xerox quality policy

Leadership through quality is founded on the Xerox quality policy, a policy written and agreed by the entire senior management team. The policy statement is as follows:

> Xerox is a quality company. Quality is the basic business principle for Xerox. Quality means providing our external and internal customers with innovative products and services that fully satisfy their requirements. Quality improvement is the job of every Xerox employee.

It interprets this as 'conforming to customer requirements'. Employees must identify their customers, which will not for the most part be the end-user of products and services, but persons within the business who use the results of the task carried out. Customer requirements were becoming more and more demanding and Xerox staff all needed to be equal to the task of meeting the new challenges. The last line of the policy statement makes it plain that improving quality is the responsibility of every Xerox employee. Before going on to look at the implications and implementation issues of the leadership-through-quality initiative we will look at the objectives defined by the Xerox management in pursuit of its policy.

The management set some precise objectives for what they wanted to achieve out of implementing leadership through quality. These are quoted below.

> To instil quality as a basic business principle in Xerox and to ensure that quality improvement became the job of every Xerox person.
>
> To ensure that Xerox people, individually and collectively, provided external and internal customers with innovative products and services that fully satisfy their existing and latent requirements.
>
> To establish, as a way of life, management and work processes that enable all Xerox people to continuously pursue quality improvement in meeting customer requirements.

3.3.6   Changing the culture

Implementation of the plan for leadership through quality was regarded by Xerox as the most significant initiative it had ever embarked on. Leadership through quality aimed to fundamentally change the organisational culture of Xerox over a relatively small number of years. It meant moving the business from the position where there was sometimes only a vague understanding of customer requirements to one where understanding and satisfying their requirements was the minimum standard; moving from accepting a margin of error, followed by corrective action, to doing things right first time, and moving away from individual action on problems to a team-work approach based on logic and consistency.

The initiative comprised three distinct but related elements: quality principles, quality tools and management behaviour and actions.

Setting the minds of Xerox staff upon leadership through quality was based on four key principles as quoted below:

> We need to understand customers existing and latent requirements.
>
> We desire to provide all our external and internal customers with products and services that meet their requirements.
>
> Employee involvement, through participation problem solving, is essential to improve quality.
>
> Error-free work is the most cost effective way to improve quality.

In order to implement the principles, appropriate tools are needed. Xerox identified five critical tools, including training:

- competitive bench-marking and quality goal setting, as touched on above;

- systematic defect and error prevention methods;

- costs of quality measurement methods;

- communications and recognition programmes;
- training for leadership through quality.

The training programme represented the largest single training investment that Xerox had ever made. The programme was for all Xerox people. The training was focused on family groups and covered the processes and tools of leadership through quality. The programme was supported by professional trainers, but the ownership was management's.

In order to make the most of the training, Xerox worked hard to ensure that the new knowledge and skills acquired by its staff were applied in a supportive environment. Communications and management behaviour and actions were considered to be fundamental to the success of the whole initiative. Xerox was also aware of the importance of recognition and reward of staff in fulfilling its objectives. Selection, appraisal and promotion processes had become targets for rethinking.

Management behaviour and actions were founded on commitments, including:

Assuring strategic clarity and consistency.

Providing visibly supportive management practices, commitment and leadership.

Setting quality objectives and measurable standards.

Establishing and reinforcing a management style of openness, trust, respect, patience and discipline.

Establishing an environment where each person can be responsible for quality.

3.3.7   Competitive advantages

The initiative dramatically changed the Xerox management processes, and leadership through quality provided the business with benefits including:

- the reliability of Xerox products, measured by customer reporting, improved;
- administrative processes were simplified;
- regular surveys showed increased customer satisfaction;
- the cost of quality declined in all areas.

As a result, Xerox reversed its market share erosion and the Mitcheldean manufacturing plant twice won the coveted British Quality Award.

3.4   Industrial experiences 2 – another step towards total quality

In this second example we look at the current moves being made by BAe MAD in quality management systems and in certification. (It is based on documents produced by John Whalley, Quality Assurance Director of BAe MAD.)[27]

This summer MAD will be assessed by Lloyds Register Quality Assurance (LRQA) with the aim of achieving registration against International Quality Standard ISO 9001. In order to ensure that the goods and services MAD provides satisfy our customers, the Division operates a Quality Management System (QMS). This consists of a number of elements: written documentation such as procedures and work instructions, management reviews and all our quality planning activities. The system provides a mechanism for implementing the key elements of our quality policy and strategy.

A successful QMS is a management system which enables everyone in the organisation to ensure customer requirements are understood and met first time, every time at minimum cost to the business. It embraces all areas of the organisation – marketing, contract acceptance, product design, production, delivery, service, finance and administration – ensuring the goods and services received by our customers are to the required standard.

Additionally, it establishes control over all our business processes, and is flexible so that it evolves to meet the changing needs of our customers and our changing business goals.

To aid the development of quality management systems generally, standards have been created that provide a check-list of the system's main ingredients. These allow the QMS to be assessed by external agencies for compliance with the standard.

Until recently MAD was assessed against the MOD-sponsored Allied Quality Assurance Procedures (AQAPs). This was known as second-party assessment – MAD being the first party and our customer MOD being the second party. Three years ago, MOD took the decision to run down its own assessment capability and instead require suppliers to be assessed against the ISO 9001 standard by an independent third party.

During the last full assessment of MAD by MOD in late 1992, the new ISO standard was used in place of the old AQAP 1. One advantage is that ISO is universally applied across a wide range of businesses – you may have seen other companies displaying their registration in their literature.

In future, only businesses that have achieved the appropriate registration will be allowed to bid for Government work. It is essential, therefore, that MAD fully meets the requirements of the standard. Registration to ISO 9001 is now a basic requirement of other nations within the European Union and is also being adopted by the United States. Achieving registration is not a one-off event – the assessors return on a regular basis to make sure standards are being maintained and improvements are being introduced to reflect changes in business requirements.

All assessment organisations are approved by the National Accreditation Council for Certification Bodies (NACCB) which, on behalf of the Government, ensures that certification bodies are competent to carry out assessments.

NACCB provides lists of certification bodies operating in each business sector and BAe MAD invited a number of NACCB approved companies to bid. LRQA were chosen using the normal commercial selection procedure. They are one of the leading assessment bodies in the UK and have a strong international presence. They have established a leading market position in the aerospace sector and include other BAe Group companies on their client list.

A great deal of work has already been done in making sure that our QMS is up to date and reflects the current business needs and requirements of the standard. With the major reorganisations that have taken place over the past few years, and the move towards a divisional structure, a lot of emphasis has been put on bringing the three previously separate systems at each unit into a single MAD system.

The opportunity has also been taken to revisit the quality policy and strategy. What is really important – not just for the assessment but also for our everyday business – is that everyone in the organisation understands our policy and strategy and those elements of the QMS that impact on their area.

An awareness drive has been launched across BAe MAD, but it requires the support of everyone in the organisation. Remember, the assessment will cover both the QMS documentation and how the system is actually being applied.

Achievement of ISO 9001 will not be the end of our quality journey. We must seek to continuously improve the quality of our products and services and the processes within our organisation. Registration, however, will be another step along the way and will be further recognition of our achievement to date.

3.5   Industrial experiences 3 – the Rover quality system

In this third example we look at the Rover Group quality system as described in a document produced by Joe Cullen, Director of Operations Strategy, Rover Group Limited.[28]

Authority for the quality system is the Rover Group Quality Council (RGQC) – this has met every quarter since 1987 and consists of the executive members of the Rover Group Board.

3.5.1   Deployment of quality responsibility

All functions and business units are responsible for ensuring that the policies established in the Rover Group Quality Policy Strategy are achieved. To this end all functions and business units have their own quality councils, which normally include the departmental heads in the area.

3.5.2   Quality in new product introduction

The basic organisation for new product introduction within the Rover Group is the project team. Fundamentally, all responsibility for the project rests with the team led by a project director. There is no independent audit of the quality of the product or process which overrides the authority of the project director. However, teams are required to work to the Rover Group Quality Policy and Strategy requirements.

The Rover Group Quality Policy requires all projects to be conducted in accordance with the Programme Management Policy (PMP). The PMP defines the process for introducing a new vehicle from business plan to volume production as a series of stages. For each stage the necessary activities are defined and the criteria for declaring the stage complete are set out in check lists. Key documents cover cost management, reliability, timing, problem and release management, environmental management and design methodology.

3.5.3   Quality in manufacture

Quality in manufacture is the responsibility of the manufacturing directors in each business unit. The manufacturing directors from across Rover Group meet regularly in the manufacturing director's forum, chaired by the company manufacturing strategy manager from Operations Strategy, to deal with common issues. There is an independent audit of finished vehicle quality carried out by Quality Audit, in the Vehicle Business Unit. The procedures and standards for audit are agreed across the company.

3.5.4   Parts quality

Rover Group's requirements from suppliers are set out in the series of documents RG2000. Rover policy is that the supplier is fully responsible for maintaining and demonstrating the quality of their product during development and production. Suppliers are responsible, contractually, for all warranty costs resulting from failure of their components during the vehicle warranty period. The purpose of this is not to penalise the supplier but to emphasise a community of interest and to encourage root-cause problem solving. Continuing supplier development is the responsibility of the purchasing department. As appropriate, supplier development teams assist suppliers to improve their processes.

Rover Group do not carry out an inspection of goods received either for first-offs or for series production. Where appropriate a supplier may be required to supply evidence of inspection results.

3.5.5   Quality of service

As with all motor manufacturers the main contact with Rover Group customers is through the franchised dealer network. Policies for customer service are set out on the Rover Group Quality Policy and are developed in more detail through the functional and Regional Sales Company organisations. Targets for improving customer service, rated as comparisons with competitors, are set out on the Rover Group Quality Strategy.

4 External supplier quality

Most manufacturers buy in two thirds by value of their final product, i.e. they only manufacture about one third of the product and assemble the whole. Therefore, when a company embarks on a quality-improvement drive, it must pay particular attention to its suppliers which may account for at least two thirds of its turnover.

Many companies have failed to develop a supplier strategy and traditionally have used an almost gladiatorial and hostile approach to their suppliers, e.g. drive them to impossibly low prices regardless of quality, or terminate trading with those who fail to perform (Figure 4.1). Most buyers are only interested in price and delivery and not the quality of the

- Gladiatorial, arms length approach to suppliers

- Buyers screw down prices

- Buyers not interested in quality or suppliers' problems

- If suppliers let us down, sack them

- We dual/triple source and play one off against another

Fig. 4.1 Gladiatorial approach to suppliers

- Price dependent

- Multi-sourcing

- Hostile

- Expediting

- Inspection

- High stocks (safety)

Fig. 4.2 A traditional approach to purchasing

supplier's goods and services. The main characteristics of a traditional approach to purchasing are shown in Figure 4.2.

The more enlightened companies have now established a procurement strategy which controls the quality of a supplier but does not take from a supplier the responsibility for his quality. The company will collaborate with the supplier through the product life cycle from specification through design, test and manufacture to ensure standards are maintained throughout the delivery and provide in-service support as shown below:

- choose suppliers who produce the required quality, not who offer the lowest price;

- reduce number of suppliers;

- build up close working relationships, reduce costs, two teams working together;

- provide collaboration and mutual trust;

- ensure operational integration at all stages of the design and build process.

This approach has demonstrated a marked reduction in supplier prices, improved quality and delivery and external customer satisfaction. The supplier must be regarded as part of the team and his full commitment to the project will ensure minimum inspection on receipt and the implementation of joint improvement programmes. More on the benefits of the approach is found elsewhere.[29,30]

4.1 Procurement strategy

An improved procurement strategy is shown below:

- reduce supplier base;

- develop suppliers as part of the team;

- ensure supplier commitment to contract;

- improve supplier delivery and quality performance;

- change supplier monitoring criteria;

- reduce receiving inspection;

- introduce product-assured suppliers;

- engage in joint improvement programmes.

A company will usually categorise its suppliers according to their type of operation. For example a company may:

- fully design, develop and manufacture its own products;

- manufacture major components from supplier designs;

- manufacture simple components from third-party designs.

In every case the company may demand that the supplier be registered to BS EN ISO 9000 or conform to their company requirements for the adequacy of their quality systems. However, as previously stated these registrations do not control the quality of the product itself. Some large companies have now introduced a product control system based on the following elements:

- the quality of the product on delivery, i.e. conformance, packaging, damage, maintenance of standards, etc.;

- the quality of the product from assembly to final delivery;

- the quality of goods in service;

- delivery to schedule and responsiveness to change;

- continuous quality improvement with resulting reductions in price.

Any new suppliers must show evidence of their capability with other contracts before being awarded any orders.

Any current supplier who fails to meet the standards will be removed from the 'bidding list'!

The engineering department of a company usually works alongside the quality-assurance department ensuring adequate specification, design, test and validation stages are carried out by the supplier. The importance of the specification cannot be overemphasised. It should be possible to trace by audit, from the final product specification to the most minor supplier products. The specification hierarchy is shown in Figure 4.3 and the supplier life cycle in Figure 4.4.

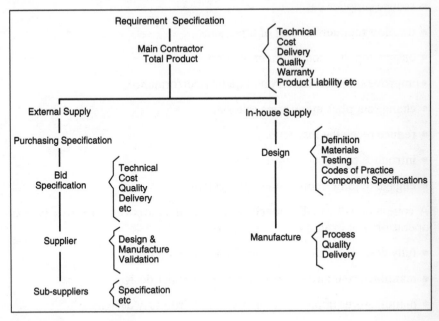

Fig. 4.3 Specification hierarchy

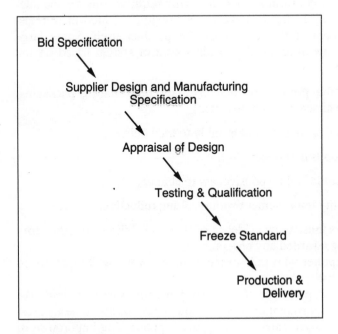

Fig. 4.4 Supplier life cycle

Finally, in parallel with quality and system approval, commercial and financial backgrounds will be investigated before the award of any orders.

The following case study is taken from a BAe MAD document[31] used to assess and improve its supplier base and develop better relationships through support and cooperation with its suppliers. The document is worded as if the reader is the supplier to BAe MAD.

4.2 Industrial experience – preferred-supplier process

4.2.1 What is the preferred-supplier process?

This is all about improving performance together. The preferred-supplier process is a continuous cycle of improvement that encompasses your entire business operation. Working with you, the supplier, we identify key opportunities for improvement and jointly agree on levels of achievement towards which you should work (see Figure 4.5).

There are three elements against which you will be assessed:

- statistical process control;

- a detailed business assessment;

- measurement of your performance.

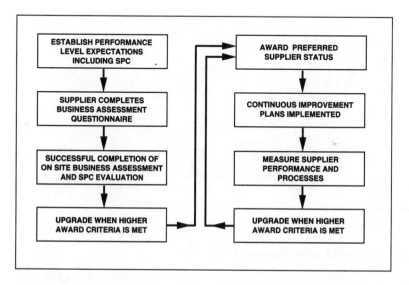

Fig. 4.5 Preferred supplier process

You must attain at least bronze standard in each category before you become a preferred supplier.

Preferred suppliers will be awarded certification. By working towards ever higher standards through a process of continuous improvement, a bronze can become a silver and a silver can become a gold (see Figure 4.6).

Some relationships have to be worked at – and customer/supplier relationships are no exception. The preferred-supplier process is designed to promote a good working relationship that drives continuous improvement from both sides – ensuring top quality performance and products.

Of course, the benefits of the process are two way. We also want our suppliers to gain from their involvement in the process. By achieving preferred-supplier status, you should expect:

- improved product quality;
- opportunities for cost reduction;
- increased business visibility;
- involvement at the design stage if appropriate;
- communication and sharing of best practice;
- enhanced customer satisfaction.

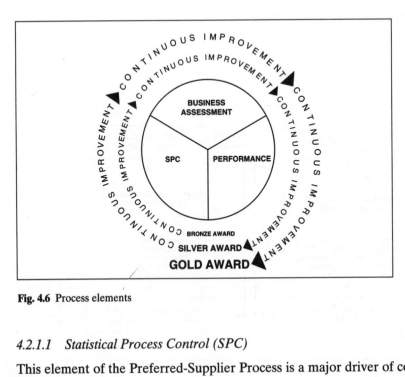

Fig. 4.6 Process elements

4.2.1.1   Statistical Process Control (SPC)

This element of the Preferred-Supplier Process is a major driver of continuous improvement.

SPC provides a structured environment which continuously reduces variation in product quality by concentrating effort on prevention rather than detection. In other words, it gets to the root of the problem (see Figure 4.7).

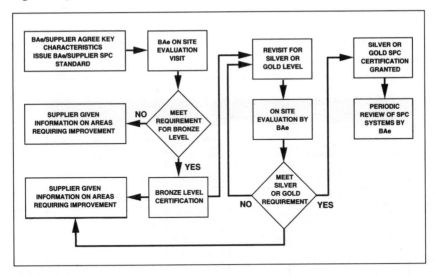

Fig. 4.7 SPC process

The adoption of SPC and its associated techniques have been shown to provide an effective way of producing higher-quality products at lower cost.

We at BAe MAD are committed to establishing positive programmes with our suppliers to ensure an effective implementation of SPC in our suppliers' organisations to complement the introduction of SPC within BAe MAD.

If SPC is completely new to you – don't panic. Our minimum requirement is for a supplier to have an implementation plan and an SPC 'pilot' with agreed characteristics identified and in place. Your plan should address all critical manufacturing processes expanding into business and administrative practices.

4.2.1.2   Business assessment

Each supplier's business processes are put under the spotlight by a team of assessors from BAe MAD, who will typically spend a week visiting your business and meeting your people. The assessments are highly detailed and the team rates how well a supplier is performing in six key areas:

- management;
- quality;

- delivery;

- cost;

- technology;

- support.

The business assessment categories and their sub-elements are shown in Figure 4.8.

MANAGEMENT	QUALITY	DELIVERY	COST	TECHNOLOGY	SUPPORT
ORGANISATION	QUALITY POLICIES AND PROCEDURES	DELIVERY SYSTEM INTEGRATION	LABOUR AND MATERIALS	PRODUCT DEVELOPMENT	ORGANISATION
PERSONNEL MANAGEMENT	PROCESS CONTROL	ON TIME MANUFACTURING DELIVERY	RATES ACCOUNTING AND	PRODUCT DESIGN DEFINITION	LOGISTICS SUPPORT DEVELOPMENT
CUSTOMER SATISFACTION	PROBLEM PREVENTION	SYSTEMS	OTHER COSTS	MANUFACTURING EQUIPMENT	FIELD MAINTENANCE SUPPORT
BUSINESS INTEGRATION	DETECTION AND CORRECTION	PACKAGING SHIPPING AND RECEIVING	PERFORMANCE PROCESSES AND COMMUNICATION	TEST AND REWORK	
SUPPLIER MANAGEMENT	SUPPLIER QUALITY PROCESS		PROPOSAL AND PRICING		
			FINANCIAL CONDITION		

Fig. 4.8 Business assessments and sub-elements

The assessment identifies your strengths and opportunities and via a continuous improvement plan helps build long-term, mutually beneficial relationships.

4.2.1.3 *Performance measurement*

Of course, we need some accurate measures, and BAe MAD will continuously monitor each supplier's progress against the agreed metrics. We will be looking at three key areas:

- delivery – schedule adherence;

- quality – right first time;

- responsiveness – assessed and reported using inputs from the people at BAe MAD who deal with you, the supplier, on a regular basis (see Table 4.1).

As you might expect with such a high-profile process, our measurement criteria are challenging. Suppliers must achieve and maintain the required

Table 4.1 Performance standards

Certification level	Performance			
	Quality		Delivery	
	Acceptance per cent	Period months	Acceptance per cent	Period months
Gold	99–100	12	99–100	12
Silver	97–98	12	95–98	6
Bronze	95–96	12	90–94	6

1. Also requires acceptable responsiveness assessment rating.
2. Performance guidelines may vary with product complexity.

standards to obtain preferred-supplier certification. Reports will be issued periodically and they will provide a valuable communication opportunity between BAe MAD and yourselves.

4.2.2 Summary

This introduction provides an overview of the processes and the benefits that may be gained by working together in an environment of continuous improvement.

BAe MAD performance is largely reliant on the performance of our suppliers. It is a long-term plan and the benefits will not come overnight, but by becoming a preferred supplier to BAe MAD you will help set new standards for us – and for yourselves.

BAe MAD acknowledges the help of MacDonald Douglas Aerospace-East in the preparation of our approach to the preferred-supplier process.

5 Software quality

The importance of control and quality of software has only recently been recognised by industry. Some 30 years ago only the major organisations utilised either large, powerful (by their standards) mainframe digital computers or analogue computers. Nowadays advanced machines of enormous capacity and speed are everyday tools.

Many companies are totally reliant on their central and distributive computing for the successful operation of their businesses, for example:

- Computer-Aided Design (CAD);

- Computer-Aided Manufacture (CAM);

- Numerical Controlled machines (NC);

- factory planning;

- stock control;

- financial control;

- salary and pension control, etc.

Equally, many modern products rely on software for their operation. Whilst this may be obvious in the defence and aerospace industries, the modern high-performance car has a level of electronics for its operation comparable to the early US Mercury manned space capsules.

It is essential that a company recognises the importance of high-quality computer software and hardware and provides the necessary quality control systems.

Whilst the software is not visible as a fabricated product, the principle of software Quality Assurance (QA) is very similar. For instance, software QA can equally be defined as:

- conformance to requirements;
- fitness for purpose;
- value for money;
- customer satisfaction;
- reliability and freedom from defects.

However, software quality is more difficult to grasp than mechanical quality. The specification for a mechanical component may be a drawing with the relevant details. Software can consist of hundreds of machine instructions for a computer to perform a function. The computer requires precise instructions for its operation. Verifying whether software meets its requirements is generally far more difficult than mechanical quality. Presently, software conformance can typically only be achieved by testing during the development and in the verification stage. In large software packages the required testing is enormous and unfortunately errors may only be discovered during usage.

To establish the required quality of software we generally consider the whole of the software life cycle:

- concept;
- requirements;
- design and analysis;
- coding;
- testing;
- acceptance.

A full life cycle is defined in Figure 5.1.

A common failing of software, causing cost and overrun time scales, is the poor definition of requirements and specifications. Full specification, documentation and configuration control are essential at all stages of the development.

The selection, development and implementation of software by users usually follows the process of:

- assessment;
- education of staff;
- selection;
- installation;
- evaluation.

To ensure that the software is correctly implemented, it is vital that appropriate 'codes of practice' are defined and used.

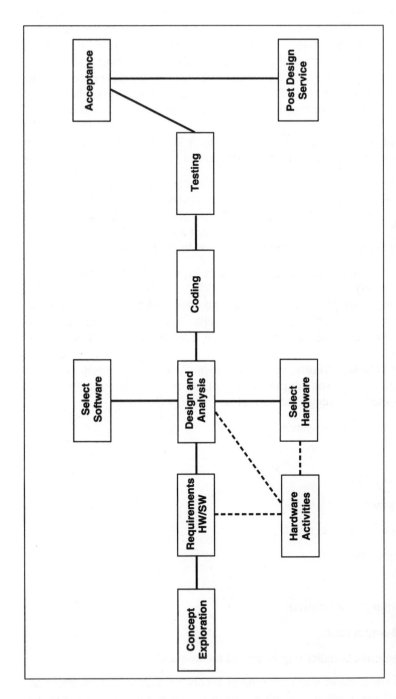

Fig. 5.1 Software life cycle

5.1   Codes of practice

The codes of practice for the introduction of new software to production are illustrated in Figure 5.2.

When a project is initiated, it is necessary to produce both a project plan and a quality plan. The quality plan must include performance objectives for the standard of the quality, a means of ensuring configuration control and how it must be tested. The codes of practice ensure that the quality plan is reviewed at appropriate intervals in conjunction with the project plan.

The concept of levels of software maturity is now being established to identify the status or ranking of the software:

1 INITIAL   –   Unpredictable and poorly controlled
2 REPEATABLE   –   Can repeat previously mastered tasks
3 DEFINED   –   Process characterised, fairly well understood
4 MANAGED   –   Process measured and controlled
5 OPTIMISED   –   Focus on process improvement.

These levels of maturity are represented in Figure 5.3 and are amplified in Table 5.1 (p. 50).

The standard of quality is that which the operation and user require, i.e. we would expect a much higher level of reliability, integrity and error-free operation for, say, a nuclear power station than that for the stock control of a factory.

Software Quality Assurance follows a similar format to any product or service with perhaps the emphasis on verification and validation and product acceptance. The suggested format is:

- quality management system;

- contractual aspects;

- quality objectives;

- quality planning;

- configuration management;

- design reviews;

- audit;

- verification and validation;

- product acceptance;

- development of quality improvement techniques.

The software reviews and audits form a very important part in ensuring quality and again follow a similar pattern to any product (see Figures 5.4 and 5.5).

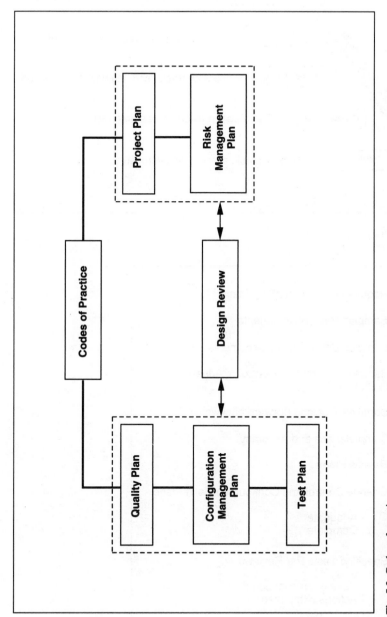

Fig. 5.2 Codes of practice

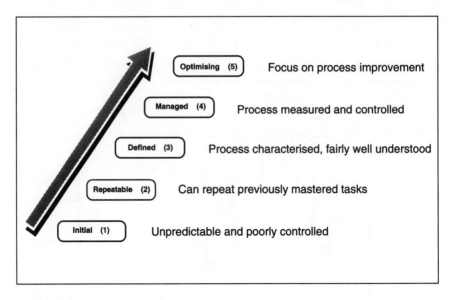

Fig. 5.3 Software process maturity

- Requirements Traceability Matrix

- Samples Review Documents

- Samples Development Documents

- Samples Design and Control of Adherence to Standard

- Samples Tracking (forward/backward)

- Completeness and Accuracy

- Review Plans

- Change Control and Configuration is in Place

    - Adequate
    - Complete

- Specified Tests Per Reviews

    - Have been conducted
    - Actions completed

- Tailoring of SQA Procedures to Project

Fig. 5.4 Software quality-assurance review

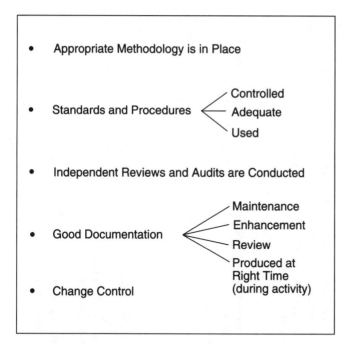

Fig. 5.5 Software audit and review

Many operations fail to establish comprehensive configuration control and produce changes in software that completely alter the function of the software or its compatibility with hardware or linked software without correct identification. Software is the same as any other product; any alteration in 'fit, form or function' transgresses interchangeability and requires changed identification (say part number).

In addition to the recent 'TickIT' certification which has been introduced into industry, there are several standards relating to the provision of software quality, including the following.

- For a software producer, ISO 9001 must be interpreted according to ISO 9000-3 guidelines. The only meaningful certification is ISO 9001/9000-3.

- BS 7165: 1991 'Recommendations for achievement of quality in software'.

Further information on the application of ISO 9001 in the software world can be found elsewhere.[32,33] The ISO 9001/9000-3 coverage of the software development process is shown in Table 5.2.

It is interesting to note that as well as the total-quality issues touched on above, other quality-improvement approaches including quality circles, and quality tools and techniques such as Pareto analysis, cause-and-effect (Ishikawa) diagrams, quality function deployment and statistical methods,

Table 5.1 Software maturity ranking

Maturity level	Key process areas	Organisation characteristics	Process capability
1. Initial		Inconsistent management attention, ad-hoc, intuitive process	Unpredictable cost, schedule and quality performance
2. Repeatable	Software subcontract management Software Quality Assurance Software configuration management Project planning Project management	Management oversight of project Project plans and status known Stable product base lines Inconsistent process	Reasonable control of schedules, but informal and occasionally ad-hoc process
3. Defined	Organise process focus Training programme Software engineering interfaces Software requirements analysis Software design Software testing Peer review	Stable and consistent process Personnel trained and follow a defined process	Reliable costs and schedule performance, improving but unpredictable quality
4. Managed	Quality management Process measurement and analysis	Product quality planning and tracking Measured and controlled software process	Reasonable statistical control over product quality but not most cost efficient
5. Optimising	Defect prevention Process improvement	Continuous process improvement	Quantitative basis for capital investment improving process efficiency

Table 5.2 ISO 9001/9000-3 coverage of software development processes

Process	ISO 9001	ISO 9000-3
Measurement	Not covered	General guidelines on the measurement of product and process
Requirements management	High level (two paragraphs)	Limited to contractual approach
Project management	High level (one paragraph)	Basic elements
Subcontractor management	Product purchase view	Basic guidelines for software development subcontracts
Quality system	Heavy coverage, but in terms of conformance to specifications, requirements and documented processes	Guidelines for software quality planning
Design	Not covered	Recommendation that adequate methods and tools be used
Implementation	Not covered	Recommendation that adequate tools be used
Verification and validation	Basic design reviews	Basic guidelines on testing and acceptance testing
Configuration management	Basic document control	Fundamentals of source-code configuration management
Maintenance and support	Not covered	Basic guidelines on corrective and adaptive maintenance

are being advocated by the software community.[34-36] Information on quality tools and techniques can be found in Chapter 6.

More and more it appears that mechanical or hardware quality and software quality are very closely related. Of course there are differences. Unlike hardware, software is not subject to 'wear-out' and software quality concerns are directed to delivery condition. Also, we are obviously looking at a situation akin to product development rather than manufacturing.

5.2   Industrial experiences 1 – total quality

In this case study we briefly overview how a 200-strong software house developing a range of tailored applications and providing management and quality management consultancy embarked on developing a total quality management system and the lessons learnt. The case study is taken from the paper by Richard Francis published in the *Software Quality Journal.*[37] The business had a quality-management system certified to ISO 9001 and registered under the TickIT scheme. The company realised that ISO 9001 was only one step on the route to total quality.

An assessment of a QMS developed to fit to ISO 9001 revealed a number of limitations:

Take up within all parts of an organization remains patchy and is disconnected from the business issues

Emphasis is conformance rather than effectiveness and improvement

Concentration is on functions, tasks and products rather than processes.

5.2.1   The TQM system

With little available literature on TQM in IT and following difficulties with the definitions, the company designed a TQM system having the following elements:

Focus on client satisfaction

Cascading job objectives flowing from the board

Underpinning of process improvement

Customer satisfaction.

The first step was finding out how the business's products and services were regarded by their clients and identifying the areas where there was scope for improvement. A client survey was used to produce subjective measures of software quality. Quality-improvement teams were formed and focused on the specific issues identified and their improvement targets.

5.2.2   Job objectives

The Board drew up measurable objectives critical to the success of the business. This set of objectives formed the first level and basis for a cascade of objectives for delegation throughout the business. In this way each member of staff would have objectives linked to the success of the company and could see more clearly the role they were playing. A survey of staff satisfaction was used to support the identification of areas for improvement.

5.2.3   Process improvement

A set of key processes for delivery of the business's critical success factors was developed to underpin the TQM system. This required a departure from thinking along functional lines (project managers, line managers and account managers) and setting up the organisation from a process perspective. This area proved to be the most challenging. The business needed to identify the key processes and process owners, and develop ways of representing processes and setting process improvement targets.

5.2.4   Lessons learnt

Expectations of the company's staff must be engineered from the outset. It is too easy to be overambitious, particularly when new ground is being broken

Continuous process improvement is difficult with the current level of poor measurement tools and techniques

Implementation must be tackled bottom-up as well as top-down. This helps prove and substantiate high level concepts and can generate early, small successes in implementation

Allow engineering judgment to help; not everything can be objectively stated and measured

Most of the development process is creative; this limits the capacity for process performers to prescribe improvements.

5.3   Industrial experiences 2 – trends in pursuit of quality

Under this heading, the results of two studies[38,39] concerned with software development trends including the pursuit of quality are outlined. It touches on defect-removal rates and gives a view on ISO 9000 in the software industry. The notes presented here are based on the article by Angela Burgess published in *IEEE Software* in 1994.[40]

A wide range of quality levels were discovered.[38] AT&T and IBM are said to average 95% cumulative defect-removal efficiency, and the software for the Patriot and Tomahawk missiles was also put in the 95%

defect-removal range. However, few MIS producers go above 75% defect-removal efficiency, although data was scarce.

The Business Practices Survey[39] discovered that quality assurance is popular at the moment. More than 50% of the respondees reported having a formal quality-assurance programme. Of the companies responding, 31% had between one and five employees only. The survey found that businesses of all sizes had a quality-assurance administrator, who was normally of middle-management status. Such administrators were more common in the larger companies. Regular formal inspections, especially of requirements, were said to be conducted by 57% of the businesses.

A broad movement towards external certification was not apparent. Only 37 of the 929 companies said that they had been 'certified' against ISO 9000, the Software Engineering Institute's Capability Maturity Model or the US National Institute of Standards and Technology's Malcolm Baldrige Award.

Cost is suggested as one reason why software businesses do not go for external certification. Comments on ISO 9000 attributed to Capers Jones[38] are as follows:

ISO 9000 is expensive, a pain to go through, and a professional embarrassment

The real weakness of ISO 9000 is that it backs away from serious quality measurement and has no standard provisions for estimating or measuring defect potentials, defect removal efficiencies and delivered defects

Because it raises the cost of developing software by a tangible amount, ISO certification could actually be detrimental to competing.

These views are somewhat contentious and are not shared by the authors. They point to the problems faced by an industry in coming to terms with the need for change. Software development is a creative and intellectually intensive domain. It is in just such situations that the need for processes, quality cultures, structured design, tools and techniques and a feedforward approach is greatest, if a business is to be competitive. The ISO 9000 notions of quality planning and defect prevention have a role to play in this connection, as mentioned in Chapter 3.

6 Quality tools and techniques

Training for quality is a vital element in a company quality campaign. It is quite incredible that managers will support extensive technical, management and other training but will be indifferent to the need for quality training and will often only support initial training.

Quality training is vital for every employee within an organisation; it is not a 'once only' effort but a continuous process for all employees, usually on an annual basis.

A list of the tools and techniques touched upon in this text is given in Figure 6.1, together with a view on how they might be typically classified. Note that in application some of the tools and techniques are frequently used in an integrated way to great effect. For example, Pareto analysis is often used in conjunction with SPC to focus its deployment on the key issues. More will be said about application patterns later. Many of the tools are very straightforward in their approach and benefit from being taught in groups or teams.

6.1 Tools and techniques overview

Under this heading, some of the more commonly used quality tools and techniques are briefly introduced. A more detailed description of FMEA, SPC process capability studies, DOE, QFD and Poka Yoke can be found in Chapter 9.

6.1.1 Stakeholder analysis

This is the identification of customers and suppliers, both internal and external, and hence requirements, communications, performance levels, service agreements, etc.

A stakeholder diagram defines:

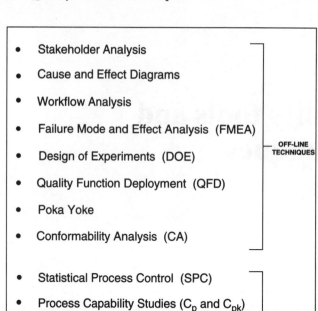

Fig. 6.1 Quality tools and techniques

- customers;

- who is involved in my success or failure;

- primary/secondary/end-user customer and highlights. Interface Management.

An example of a Stakeholder design[41] for a production function is shown in Figure 6.2. A stakeholder analysis helps to establish the customer/supplier relationships.

A customer is anyone, inside or outside the company, who receives an output from you. Customers are the only people who can really tell us how we are performing. The analogy of a company to a chain where each joint in the chain represents an internal customer/supplier interface is shown in Figure 6.3. Having identified your customers/suppliers, you then define the relationship and methods of improvement (Figure 6.4) (see also Chapter 4).

6.1.2   Pareto analysis

This approach identifies the major and vital factors in an analysis. It directs attention to the main problem where only limited resources are

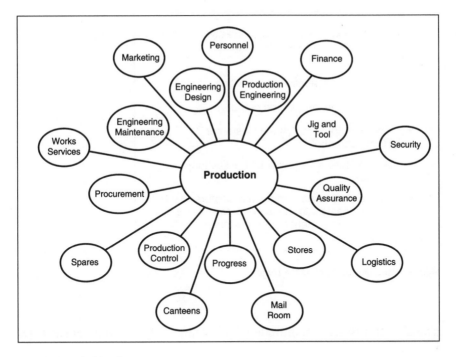

Fig. 6.2 Stakeholder diagram

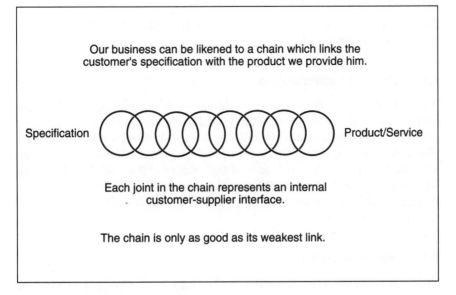

Fig. 6.3 Customer/supplier relationship

How do we take what our Suppliers give us, and turn it into what our Customer needs?

- What are the processes, activities and tasks?

- What are the inputs and outputs?

- Do the outputs match the Customer Requirements?

- Are we sure we know what the Customer wants?

- Do we choose performance indicators which are the easiest to measure, or the ones which are most important to our Customer?

Fig. 6.4 Improving your link in the chain

available to resolve a multitude of problems. A Pareto diagram enables you to:

- analyse problems from a new perspective;

- focus attention on problems in order of priority;

- compare changes in data over time;

- display problems on a commonly used format.

An example of a Pareto diagram is shown in Figure 6.5. Further information is given elsewhere.[42-44]

6.1.3   Cause-and-effect analysis

This is sometimes referred to as a fishbone or Ishikawa diagram. The diagram is a visual representation used to identify the causes of a particular problem. The diagrams are generally constructed through team effort with the advantages of group 'brain-storming' or 'lateral thinking'. They are sometimes referred to as 4Ms, where the Ms relate to Machines/Methods/Materials/Manpower, or the 4Ps of People/Products/Price/Promotions. An example is shown in Figure 6.6. Further examples are given elsewhere.[42-44]

6.1.4   Workflow analysis

This is a graphical representation of the steps in producing a product or service.[41] It is an invaluable analytical tool which can quickly identify those

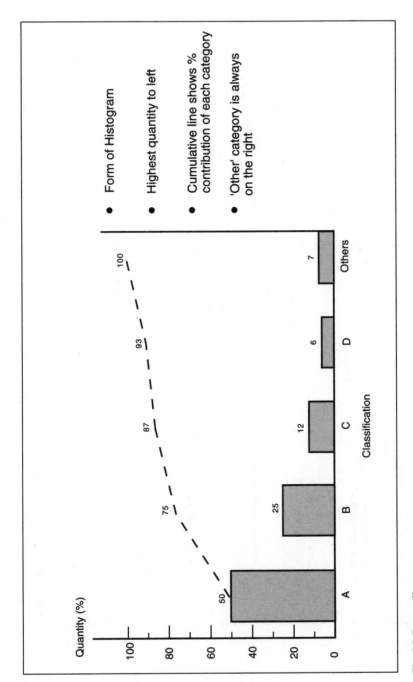

Fig. 6.5 Pareto diagram

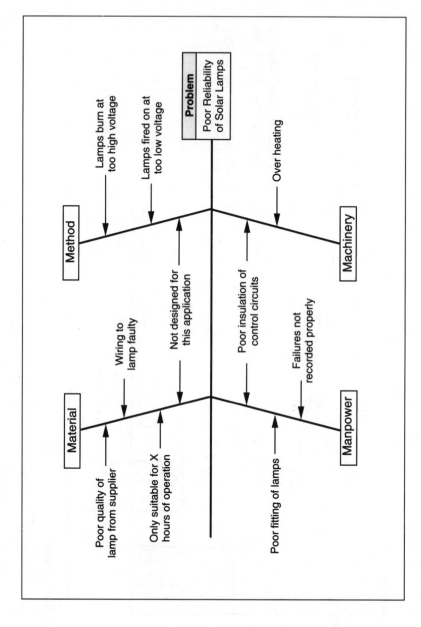

Fig. 6.6 Example of cause-and-effect diagram

activities that 'add value' to the process and those that do not. It is a means of portraying a sequence of events diagrammatically using symbols to help a person visualise a process in order to examine and improve it, i.e. a series of customer/supplier diagrams put together. After drawing out the workflow analysis chart, each part of it is examined to see if it adds value. If not, it is eliminated! What is left is simplified.

Such an analysis enables you to:

- understand the process;

- estimate quality costs;

- determine project team membership;

- plan data gathering;

- generate theories;

- identify data stratification;

- identify opportunities for streamlining;

- explain the process to others;

- change the process;

- examine each decision symbol;

- examine each rework loop;

- examine each activity symbol;

- examine each document or database.

A typical workflow chart is shown in Figure 6.7 illustrating the improved material flow through a machine tool shop after performing a workflow analysis.

6.1.5 Histograms

A process under control exhibits a minimum variation in the products. A histogram is a graphical representation of the variation, demonstrating pattern and range of variation. However, it is an analytical tool only and does not provide a remedy. It illustrates patterns that are difficult to see in a table of numbers. Common histogram patterns are given in Figure 6.8.

Major uses include:

- helping to identify root causes (by demonstrating the pattern of variation in the data);

- helping to check process performance.

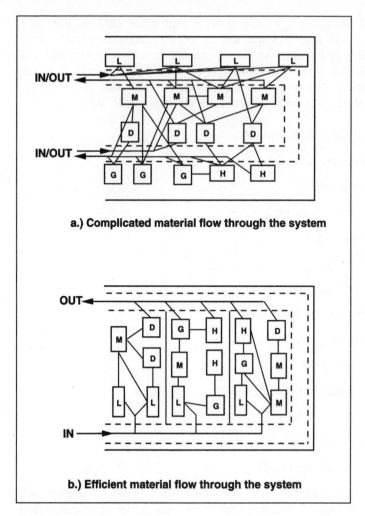

Fig. 6.7 Example of workflow chart

6.1.6   Failure Mode and Effect Analysis (FMEA)

This is an important and systematic tool generally used in the design process to identify the effects of individual element failures on the whole assembly or process. It is sometimes referred to as 'what if' failure analysis of a particular item or sequence and its effects on the assembly or process. The analysis is used in many high-technology products and is mandatory in civil aircraft design.

See Chapter 9 for a more detailed description of FMEA and worked example.

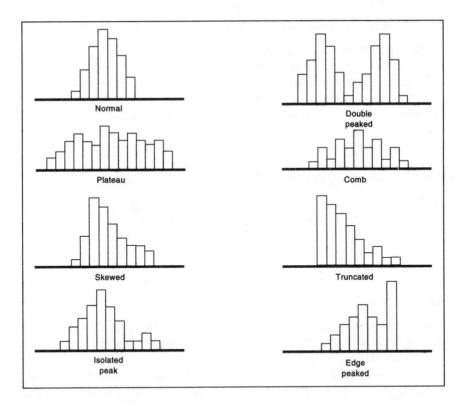

Fig. 6.8 Common histogram patterns

6.1.7   Statistical Process Control (SPC)

A powerful analytical tool initially developed in the mass production environment but now being used in batch-type production. It is used to monitor a process to detect change or variation and the need for adjustment or correction. The correction or adjustment is then fed back into the process. See Chapter 9 for more detail and worked example.

6.1.8   Process capability studies

Statistical techniques are used to measure the performance of a process. From samples of components, process capability indices are calculated and used to indicate whether a process is in control or not. The indices (C_p and C_{pk}) measure the capability of a process to produce parts consistently to meet design requirements, and as such are measures of the ability to consistently conform to specifications. See Chapter 9 for more detail and worked example.

6.1.9   Design Of Experiments (DOE)

These are methodologies for experimental design (off-line) which complement on-line quality-control systems and can be used in most areas of
product design and manufacture. Statistical methods can be used to minimise the number of changes required for each variable when looking for
an optimal solution. Considerable benefits can be achieved over the more
traditional, one-variable-at-a-time experiments. See Chapter 9 for more
detail and worked example.

6.1.10   Quality Function Deployment (QFD)

QFD supports the definition and ranking of customer requirements and
cascades them down through the product development process in four separate phases. It yields a quantification (judgemental) between product
design issues and the important customer requirements. See Chapter 9 for
more detail and worked example.

6.1.11   Poka Yoke

The provision of effective quality assurance to prevent defective products
being produced is essential. The emphasis is moved from inspecting to preventing the manufacture of any defective products. In the drive towards
zero defects, the use of Poka Yoke, or foolproofing, within all stages of the
production process is crucial and needs to be considered and implemented
wherever possible. See Chapter 9 for more detail and worked example.

6.2   Industrial experiences – tools and techniques in practice

The correct positioning of the various off-line quality tools and techniques
in the product introduction process is an important consideration. Patterns
of application have been proposed by a number of workers over several
years.[45-47] The approach presented here is that adopted by the Electronics
Division of Lucas Industries in their manufacturing design operations[48]
(see Figure 6.9). It is particularly relevant since it explicitly includes activities under the heading of DFQ concerned with quality of conformance.

The QFD process is used to help understand and quantify the importance of the customer requirements, and to support the definition of
product requirements. The FMEA process is used to explore potential
failures, their likely severity, occurrence and detectability. The Lucas DFA
(Design For Assembly) technique is used to minimise partcount and facilitate ease of assembly. DFA techniques are primarily aimed at cost reduction and have found application in many businesses worldwide. More
information on techniques in DFA can be found elsewhere.[49-51]
Conformability Analysis (CA) (see Chapter 7) is employed to provide a
measure of potential process capability in manufacture and assembly, and
to ensure robustness against failure and quality costs.

When used in conjunction with DFA, Conformability Analysis helps to

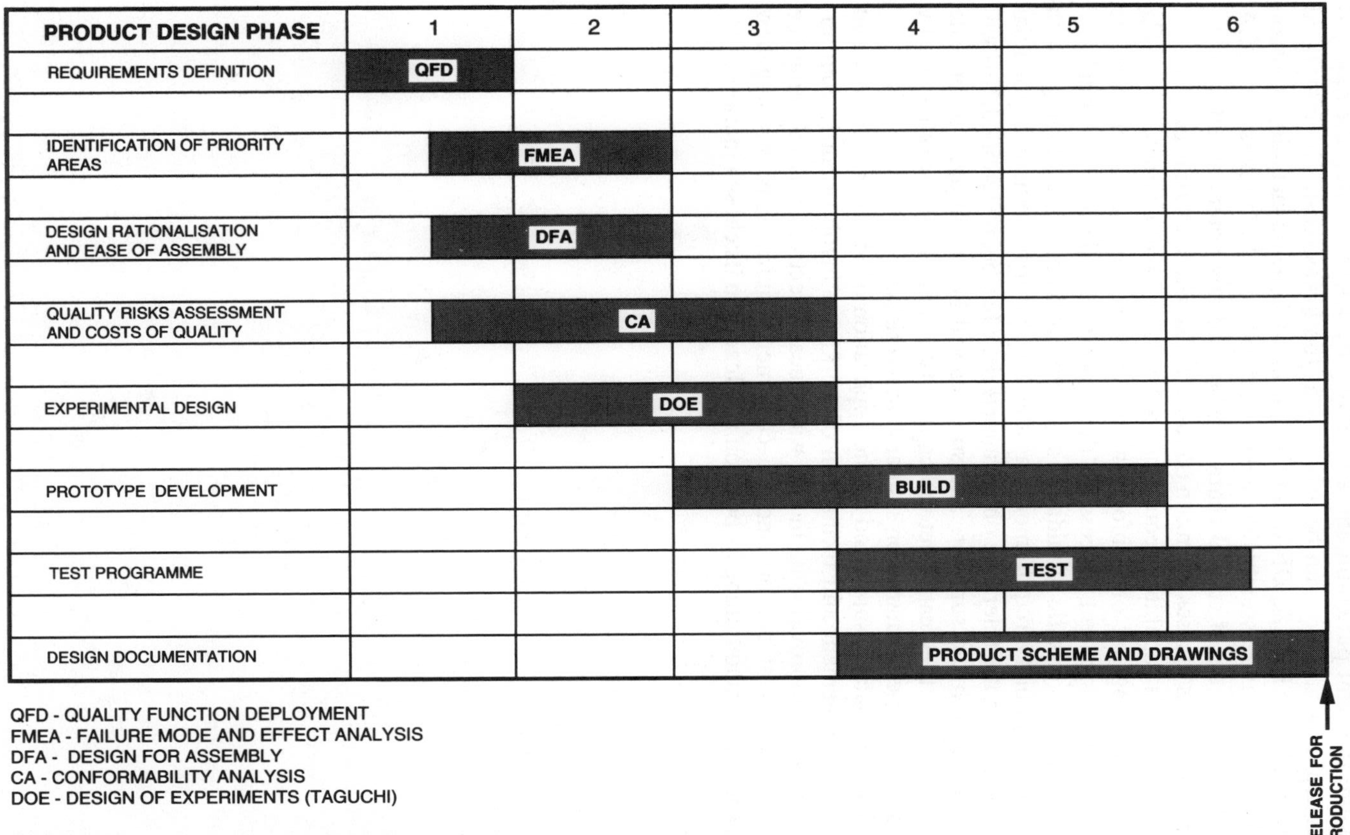

PRODUCT DESIGN PHASE	1	2	3	4	5	6
REQUIREMENTS DEFINITION	QFD					
IDENTIFICATION OF PRIORITY AREAS		FMEA				
DESIGN RATIONALISATION AND EASE OF ASSEMBLY		DFA				
QUALITY RISKS ASSESSMENT AND COSTS OF QUALITY		CA				
EXPERIMENTAL DESIGN		DOE				
PROTOTYPE DEVELOPMENT				BUILD		
TEST PROGRAMME					TEST	
DESIGN DOCUMENTATION				PRODUCT SCHEME AND DRAWINGS		

RELEASE FOR PRODUCTION

QFD - QUALITY FUNCTION DEPLOYMENT
FMEA - FAILURE MODE AND EFFECT ANALYSIS
DFA - DESIGN FOR ASSEMBLY
CA - CONFORMABILITY ANALYSIS
DOE - DESIGN OF EXPERIMENTS (TAGUCHI)

Fig. 6.9 Typical bar chart for a product introduction process

assure the right partcount and assembly port designs so that the required level of conformance can be achieved at the lowest cost. The integrated application of FMEA, DFA and CA has many benefits including common data sharing and assembly sequence declaration. The outputs of FMEA and CA in terms of critical characteristics and any potential out-of-tolerance problems focuses attention on those areas where DOE is needed.

The process capability requirements for component characteristics (C_{pk} values) coming out of Conformability Analysis are used to support the supplier development process. In the process new designs of 'bought-in' parts, castings, plastic mouldings, assembly work, etc., are discussed with process capability requirements as the first priority. Where a potential problem with the tolerance on a characteristic has been identified, this will be raised with the supplier. The supplier will be encouraged to provide evidence that they can meet the capability requirements or otherwise by reference to performance on similar characteristics.

The approach is supported by the application of SPC in the factory and the encouragement and facilitation of workforce involvement in continuous quality improvement. In this way Lucas Electronics is attaining the highest levels of conformance to customer requirements, winning new business and maintaining its position as a leader in its chosen markets.

7 Designing for quality

A quality review of a number of engineering teams over a range of business sectors identified the following problems in procedure:

- inadequate design for manufacture;

- overdesign in pursuit of excellence;

- poor specifications;

- too many design errors.

Many engineering design staff do not seem aware of their dual responsibility for quality:

- the quality of their individual work;

- the quality of the ultimate product specified by the engineering department.

In many industries, by the time the drawings/specifications/databases are issued for manufacture, approximately 5–7% of the total life-cycle costs will have been spent. However, usually about 75% of the total life-cycle costs will have been committed (see Figure 7.1). At the end of the project definition stage we have defined the means by which the product will be made, i.e. material, geometries, manufacturing route, assembly processes, etc. will have largely been predetermined. We have also established much of the costs of its in-service use, i.e. reliability, maintainability, quality, operating costs, etc. It is vital, therefore, that the engineering department's work is correct, otherwise the cost of rectification is enormous. Studies have shown that the late detection of problems can be several orders of magnitude more costly than prevention or early-stage detection.

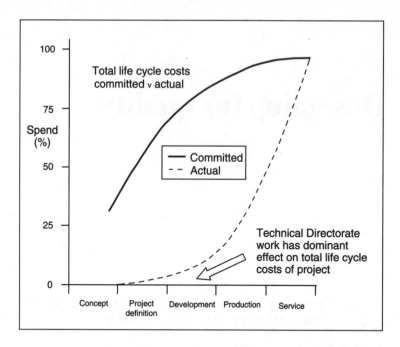

Fig. 7.1 Quality in the technical directorate

Any implementation of the concepts of 'design for quality' or 'competitive engineering' must take the following into account:

- Design reviews are a formal procedure to establish the total understanding and acceptance of a design task. A design review must not focus on cost and time scales, it forms a traceable record of technical decisions and performance against requirements. Design reviews are a crucial part of product introduction and greater details on the what, when and how of such reviews are given in Appendix 1. Feedback from various 'tools and techniques' is an important part of the design review process.

- Adequate Research and Development (R&D) and Project Definition (PD) should be completed before the launch of the project.

- Adequate specifications – the majority of the specifications, i.e. product, technical bid, materials and processes are written by the engineering department (see Chapter 4 and Figures 4.3 and 4.4).

- Supplier quality – as stated previously, the engineering department plays a major overseeing role with suppliers through design, test and qualification prior to production (see Chapter 4).

- Configuration Control, i.e. the definition of a product in terms of the drawings, specifications, software, databases and their issues which together give the configuration of the product service. (The software aspects have been discussed previously in Chapter 5.)

It has been shown that a product's cost and quality are largely predetermined during the product design process.[45,52] Also, product introduction lead times are often protracted by late engineering change. Therefore, there are limits to the extent to which competitive business performance can be improved by investment in best practice in manufacturing and shop-floor quality control.

According to the German Quality Assurance Programme for 1992–96,[53] studies show that around 75% of all faults detected in products originate during the development and planning stage, but some 80% of faults are often detected in finished parts or products, and when they are in use by the customer (Figure 7.2).

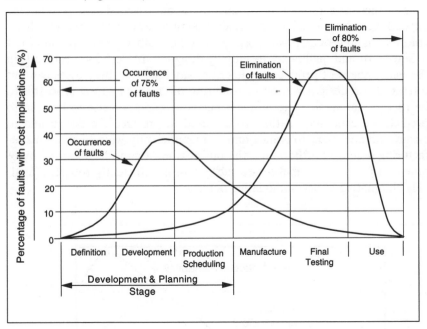

Fig. 7.2 Occurrence and elimination of faults in a product life cycle

A study of engineering change in nine major businesses in the aerospace, industrial and automotive markets[54] showed that, on average, almost 70% of product engineering rework was due to quality problems, i.e. failure to satisfy customer expectations and to anticipate production variability on the shop floor. The need for more than 40% of the rework was not identified until production commenced.

The reasons for rework were classified into four groups:

(a) customer-driven changes including technical quality;

(b) engineering science problems (stress analysis errors, etc.);

(c) manufacturing/assembly feasibility and cost problems;

(d) production variability problems.

The disposition of the rework is shown in Figure 7.3. This indicates that customer-related changes (a) occurred throughout concept design, detailing, prototyping and testing with some amendments still being required after production had begun. Engineering science problems (b), which represented less than 10% of the changes on average, were mostly cleared before production commenced. The most disturbing aspect is the acceptance by the businesses that most of the manufacturing changes (c), and more so manufacturing variability (d), were taking place during production, product testing and after release to the customer. Because the cost of change increases rapidly as production is approached and passed, the expenditure on manufacturing quality-related rework is extremely high. (More than 50% of all rework occurred in the costly elements (c) and (d).)

Since cost and quality are essentially designed into products (or not!) in the early stages of product engineering, any idea that businesses can put tolerances on designs without considering the manufacturing processes to be used is untenable. The business needs to know, or else to be able to predict, the capability of the process used to produce the design and to ensure the necessary tolerance limits are sufficiently wide to avoid manufacturing defects. Furthermore, the business must consider the severity of potential failures and make sure the design is sufficiently robust, effectively eliminating or accommodating defects.

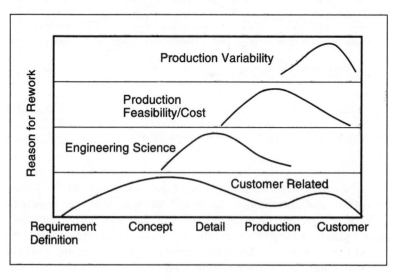

Fig. 7.3 Disposition of rework in product introduction

Given the above, it is clear that quality begins at the design stage. We will now go on to present Design For Quality (DFQ) ideas and discuss their application in product development.

7.1 Developments in design for quality

There is relatively little work published in the field of DFQ compared with, say, design for strength. This is not perhaps very surprising since the notion of DFQ and the importance of quality in designs are quite recent additions to both business and academic agendas.

A number of attempts have been made to define DFQ at various levels of abstraction, for example Hubka[55] suggests that:

Design For Quality is a knowledge system (part of the total design science) that gives all the necessary knowledge to a designer to achieve the requested 'quality' of a product or process.

A substantial review of DFQ and the framework for its application has been proposed by Mørup[56]. Mørup identifies eight key elements in DFQ which are placed under the heading of preconditions, structured product development and supporting methods/tools and techniques, as illustrated in Figure 7.4.

In thinking about DFQ, it is convenient to divide product quality into two main categories:

- 'big Q' which is customer/user perceived quality;

- 'little q' which relates to our efforts in creating big Q.

The above notions are represented diagrammatically in Figure 7.5.

Product quality is a vector with several types of quality as elements.[57,58] The Q vector relates to issues including: reputation, technology, use, distribution and replacement.

The term q can also be considered as a vector with elements related to variability in component manufacture, assembly, testing, storage, product transport and installation. The notion of q can also be expressed as an efficiency related to efforts in meeting Q.

The issues of q are met when the product meets those systems that are used to realise quality Q. The maintenance of Q relies upon the ability of a business to understand and control the variability which might be associated with the process of product realisation.

Quality in a product is not directly connected to cost. Every single Q element, Q_i has corresponding q elements that contribute to cost. Q is fundamentally connected to selling price.

The general model may be used to drive the considerations of the important issues throughout the stages of production development and in the design of individual components and assemblies. (See Figure 7.6.)

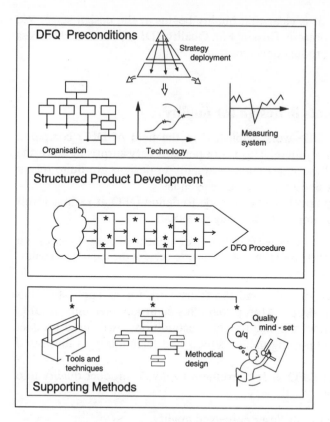

Fig. 7.4 Preconditions for and main elements of design for quality[56]

Fig. 7.5 DFQ **Q/q** concept[56]

Note that quality cannot be attached to the product alone or to the user, but arises from the interaction between the two in every product life phase.

As can be seen from the above, the DFQ and **Q/q** concepts are extremely broad in perspective. In this text we are primarily concerned with the quality of conformance aspects of DFQ and related quality costs.

7.2 Conformability analysis

The designer's job is made easier if current standard design rules are available for the process to be used to produce the design. A good set of design rules indicates process capability tolerances for a design.

Life is much more problematic when rules are not available or an unfamiliar process is to be used. The business needs to understand when required tolerances are pushing the process to the limit and to specify where capability should be measured and validated.

Conformability analysis aims to predict for a design its potential level of process capability in manufacture and assembly and to determine its robustness against failure and likely quality costs.

The provision of systematic and quantitative analysis methods and knowledge on the tolerance capability of processes, including the effects of processing non-ideal materials, geometries and applying finishing treatments, are useful in this connection. Process capability maps (graphs of tolerance against nominal dimension bounded into regions by risk level) and additional risks associated with any non-ideal characteristics can be used to quantify the likelihood of nonconformance to requirements. (A selection of process capability maps taken from the *Conformability Analysis Workbook*[59] can be found in Figure 7.7. The workbook currently contains some 60 maps covering most industrial processes.)

Recognising that the relationship between a design and its production quality is complex and not easily amenable to precise scientific formulation, the analysis has resulted largely from knowledge of engineering in manufacturing businesses, including those specialising in particular manufacturing and assembly processes. To quote: 'Engineering judgement and experience are needed to identify potential variability risks or noise associated with manufacturing and assembly routes'.[60] In the elicitation and representation of knowledge used in the analysis, the notions of single-piece, piece-to-piece and time-to-time sources of variability[61] have been included.

The quantities enumerated (**q**) correlate with shop-floor process SPC data (C_{pk}) and since C_{pk} can be related to likely fault rates, it becomes possible to predict for a design the potential number of failures due to nonconformance.[60,62]

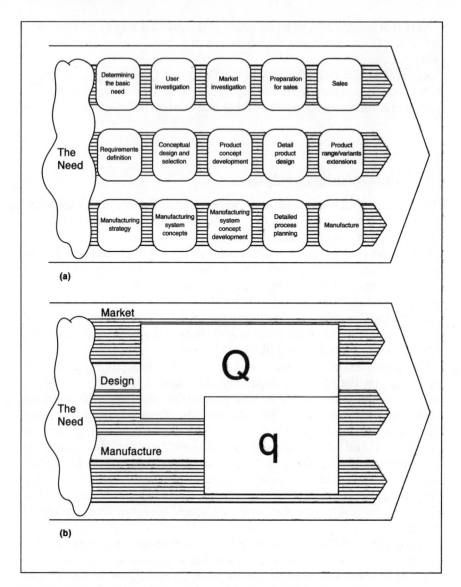

Fig. 7.6 (a) Model of integrated product development, (b) **Q/q** in the model

Here $q(q_m, q_a)$ are variability risk metrics.[59,62] Manufacturing process risk (q_m) is a function of tolerance and surface roughness to process capability, material and geometry to process compatibility, and surface engineering issues. The assembly process risk (q_a) is a function of the compatibility of component design to handling, insertion and fastening processes. For an ideal design both q_m and q_a correspond to unity. If either q_a or q_m is around 3 or above, the characteristic is unlikely to be in control.

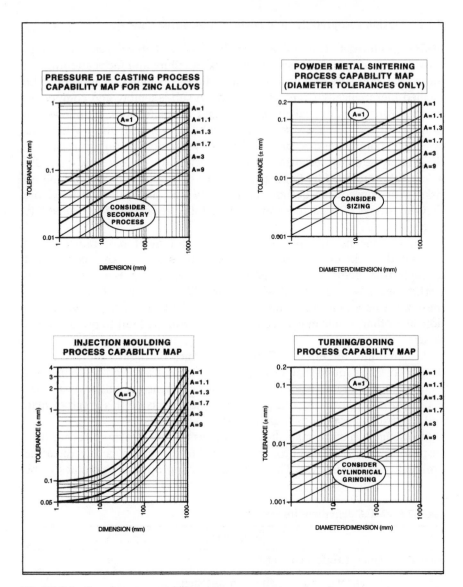

Fig. 7.7 Selection of process capability maps

C_{pk} is a process capability index[63] (see Chapter 9). If the process characteristic is a normal distribution, C_{pk} can be related to a parts-per-million (ppm) defect rate. $C_{pk} = 1.33$ equates to a defect rate of 30 ppm at the nearest limit. At $C_{pk} = 1$ the defect rate equates to 1300 ppm (see Figure 9.12).

7.2.1   Design acceptability and costs of nonconformance

FMEA (Failure Mode and Effect Analysis – see Chapter 9) can be used to provide a quantitative measure of the risk of failure and focus attention on those characteristics of a product where the risk is high. Because it can be applied hierarchically from system through sub-assembly and component levels down to individual dimensions and characteristics, it follows the progress of the design in detail. The team carrying out the FMEA lists potential failure modes and makes a judgemental rating regarding their likely severity, occurrence and detectability. It therefore provides a possible means for linking potential nonconformance (variability) risks with consequent design acceptability and associated costs. (Note that the ratings of occurrence and detectability are equated to probability levels.)

'Failure' means that performance does not meet requirements and is related back in the design FMEA to some product/component characteristic being out of the specified limits – a fault. (A typical FMEA severity scale is shown in Table 7.1.) For a failure to occur, there must be a fault, the fault must not be detected by tests and inspection, and other events may need to combine with the fault to bring about a failure.

Research into the effects of nonconformance on failure and associated quality costs has found that an area of acceptable design can be defined for a component/characteristic on a graph of occurrence versus severity (see Figure 7.8). Furthermore, it is possible to plot points on this graph or map and construct lines of equal quality cost (%) (isocosts).[62]

In essence, what the map shows is that as failures get more severe they cost more, so the only approach available to a business is to reduce the probability of occurrence. The map enables appropriate C_{pk} values to be

Table 7.1 FMEA definitions for severity rating

	Severity (S)
Rating	Guide to effect on user
1*	No effect or minimal effect on customer
2*	Minor annoyance to customer
3*	Annoyance to customer but no loss of major function
4*	Possible return to manufacturer
5*	Definite return to manufacturer
6	Failure leading to violation of statutory requirement
7	Failure leading to injury or a more safety critical related problem with secondary back-up
8	Safety problem – degradation of function with possible severe injury
9	Complete failure with probable severe injury and/or loss of life
10	Catastrophic failure with high probability of loss of life

*These failures are non-safety critical

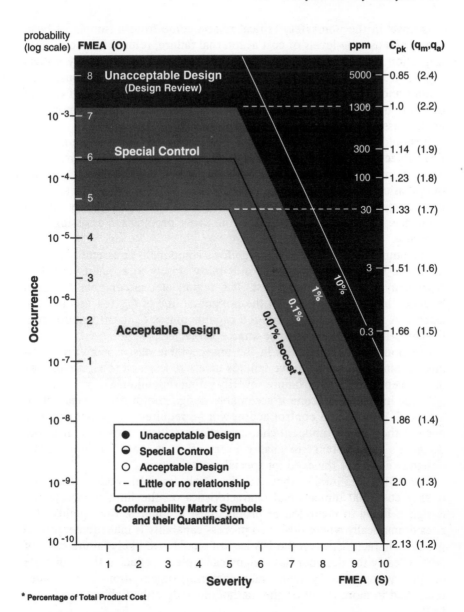

Fig. 7.8 Conformability map

selected and through linking into the conformability analysis metrics ($\mathbf{q_m}$ and $\mathbf{q_a}$ in Figure 7.8) it is quickly possible to determine if a product has characteristics that are unacceptable and, if so, what the cost consequences are likely to be.

Isocosts in the non-safety critical region come from a sample of businesses and assume levels of cost at internal failure, returns from customer inspection or test and warranty returns. The costs in the safety critical region $(S > 5)$ are based on allowances for failure investigations, legal actions and product recall, but do not include elements for loss of current or future business. The costs in the safety critical area have been more difficult to assess and have a greater margin of error.

It should be noted that losses due to safety critical failures are in practice subject to wide variation. The losses that companies face are influenced by many factors including: business market, sales/turnover, territory of operation and product liability history. It is not easy to make a satisfactory estimate of the product liability costs associated with quality of conformance. The insurance sector faces the above problems in assessing their exposure.

The boundary of acceptable design for a component or assembly characteristic in the zone $S > 5$ corresponds fairly closely to a quality cost line equivalent to 0.01% of unit cost. The region of unacceptable design is bounded by the intersection of the horizontal line of $C_{pk} = 1$ and the 1% isocost. A process is not considered capable unless $C_{pk} \{>=\} 1$ and a quality cost >1% is considered to be unacceptable.

Components/characteristics in the unacceptable design zone are virtually certain to cause expensive failures unless redesigned to an occurrence level acceptable for their failure severity rating (minimum $C_{pk} = 1.33$).

In the intermediate zone if acceptable design conformability cannot be achieved, then special control action will be required. If special action is needed then the component/characteristic is critical. However, it is the designer's responsibility to ensure every effort is made to improve the design to eliminate the need for special control action.

The 0.01% line implies that even in a well-designed product there is a quality cost: 100 dimensional characteristics on the limit of acceptable design is likely to incur 1% of product cost in failures. But, quality cost rises dramatically where design to process capability is inadequate: just 10 product characteristics on the 1% isocost would give likely failure costs of 10%. Clearly the designer has a significant role in reducing the high costs of quality reported by many manufacturing organisations. The reader interested in more detail on the methodology described above is referred to Ref. 62.

7.3 Industrial applications of conformability analysis

Under this heading a few applications are touched on to illustrate the benefits that can accrue from understanding production variability and its effects.

7.3.1 Variability prediction

A key objective of the analysis is predicting in the early stages of the product introduction/development process the likely levels of out-of-tolerance variation when in production. To illustrate what can be done, consider the examples given in Figure 7.9 which shows components with known SPC histories together with the q_m values calculated for the design characteristics and their production routes (commercial confidence precludes the inclusion of component drawings and detailed analysis). It shows evidence of the correlation between q_m values and levels of out-of-tolerance variation. The figure indicates that the analysis is capable of enabling a business to spot potential process capability problems before production commences.

7.3.2 Tolerance chain analysis

Tolerances on components that are assembled together to achieve an overall design tolerance across an assembly can be individually analysed, their potential variability predicted and their combined effect on the overall conformance determined. The analysis can be used to optimise the design through the explorations of alternative tolerances, processes and materials with the goal of minimising the risks of conformance.

7.3.3 Mechanical components – requirements definition

Where designers require tighter tolerances than the standard they must find out how this can be achieved, what secondary processes/process development is needed and what special control action is necessary to give the required level of capability. This must be validated in some way.

The variability risks analysis described previously is useful in this connection. It provides systematic questioning of a design regarding important factors that drive variability. It estimates quantities that can be related to potential C_{pk} values, and identifies those areas where redesign effort is best focused.

Such analysis is not a constraint on design. The results serve as a good basis for comparing design alternatives and setting up a dialogue with component suppliers. This helps to establish which design alternatives are most appropriate to the component application and what special controls, if any, are necessary to ensure capability. Since the vast majority of cost is built into a component in the early stages of the design process, it is advantageous to appraise the design honestly as soon as possible.

7.3.4 Electronic components – supplier characterisation and worst-case design

The situation in electronic components tends not to be so problematic. Most suppliers are more than willing to share the data they have collected

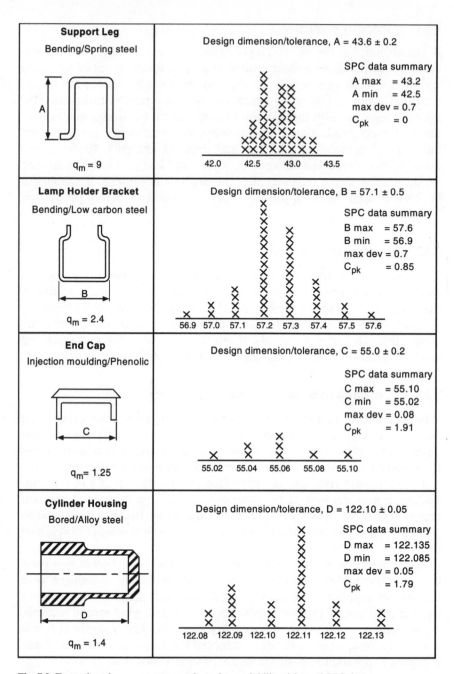

Fig. 7.9 Examples of component manufacturing variability risks and SPC data

on the capability of their processes with their customers' designers. Component printed circuit board manufacturers know what track widths and spacing they can achieve, what hole centre positions can be held and the cost trade-offs of drilling through fewer boards in a stack.

The specifications of commodity products such as electronic components must represent what the supplier can capably manufacture. It is for this reason that electronic engineers need to ensure their designs work with 'worst-case' component specifications.

7.4   Industrial example – telescopic lever assembly

The conformability analysis is based around the completion of a conformability matrix relating potential variability risk indices for component manufacture/assembly processes (in a sequence of assembly work) to potential failure modes, their severity and the costs of quality. The symbols placed in the nodes of the conformability matrix represent levels of design acceptability and are obtained from Figure 7.8.

The link with FMEA brings into play the additional dimension of potential variability into the assessment of failure modes and their effects on the customer. The method can be employed to set appropriate levels of C_{pk} for safety critical characteristics. In this way, design conformability problems can be systematically addressed, with benefits including reduced costs, improved profitability and shorter introduction times.

The conformability matrix also highlights those 'bought-in' components and/or assemblies that have been analysed and found to have conformance problems and require further communication with the supplier. This will ultimately improve the supplier development process.

To provide more insight into the application of the analysis at the component/assembly level, consider the telescopic lever assembly, design A in Figure 7.10. The assembly has an FMEA severity rating of 8, and is used in a product having a cost of £150. It is estimated that 5000 units are produced per annum.

The assembly is subjected to bending in operation, with the maximum bending stresses occurring at a point on the main tube corresponding to the stop ring recess. In order to provide compensation, a reinforcing tube is positioned as shown. This is crucial, since fracture of the telescopic lever can result in injury to users and third parties. The lower portion of Figure 7.10 shows part of the results from the conformability matrix for design A. At each node in the matrix, consideration has been given to the effect of the variable predicted from the process capability analysis (represented by q_m and q_a) on the failure mode in question (determined from FMEA analysis).

While the design is satisfactory from a design-for-strength point of view, the analysis highlights a number of areas where potential variability and failure severity combine to make the risks unacceptably high. For example, there are no design features that assure the capable positioning of the reinforcement tube in the assembly (see assembly process a3 in Figure 7.10), but there are several critical components whose tolerances need to be controlled if the tube position is to be maintained in service, such as the

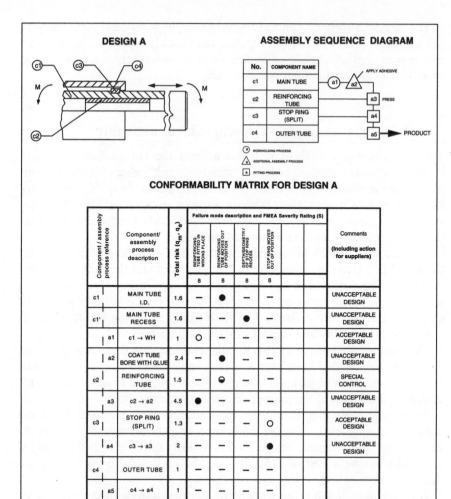

DESIGN A

ASSEMBLY SEQUENCE DIAGRAM

No.	COMPONENT NAME
c1	MAIN TUBE
c2	REINFORCING TUBE
c3	STOP RING (SPLIT)
c4	OUTER TUBE

⊙ WORKHOLDING PROCESS

△ ADDITIONAL ASSEMBLY PROCESS

☐ FITTING PROCESS

CONFORMABILITY MATRIX FOR DESIGN A

Component / assembly process reference	Component/ assembly process description	Total risk (q_m, q_a)	REINFORCING TUBE FITTED IN WRONG PLACE	REINFORCING TUBE MOVES OUT OF POSITION	DEPTH/GEOMETRY OF STOP RING RECESS	STOP RING MOVES OUT OF POSITION			Comments (Including action for suppliers)
			8	8	8	8			
c1	MAIN TUBE I.D.	1.6	—	●	—	—			UNACCEPTABLE DESIGN
c1'	MAIN TUBE RECESS	1.6	—	—	●	—			UNACCEPTABLE DESIGN
a1	c1 → WH	1	○	—	—	—			ACCEPTABLE DESIGN
a2	COAT TUBE BORE WITH GLUE	2.4	—	●	—	—			UNACCEPTABLE DESIGN
c2	REINFORCING TUBE	1.5	—	◓	—	—			SPECIAL CONTROL
a3	c2 → a2	4.5	●	—	—	—			UNACCEPTABLE DESIGN
c3	STOP RING (SPLIT)	1.3	—	—	—	○			ACCEPTABLE DESIGN
a4	c3 → a3	2	—	—	—	●			UNACCEPTABLE DESIGN
c4	OUTER TUBE	1	—	—	—	—			
a5	c4 → a4	1	—	—	—	—			
Total Failure Mode Isocost (%)			10.01	13.3	3	10.01			**TOTAL COST**
Total Failure Mode Cost			£ 75.8K	£ 99.8K*	£ 22.5K	£ 75.8K			**£2741?**

* Sample calculation of quality cost:

Number of units = 5,000

Product Cost (Pc) = £150

Total Failure Mode Isocost (%) = 3 + 10 + 0.3 = 13.3

Therefore, total failure mode cost = $\dfrac{13.3 \times 5000 \times 150}{100}$ = £99,750

Fig. 7.10 Telescopic lever assembly – design A

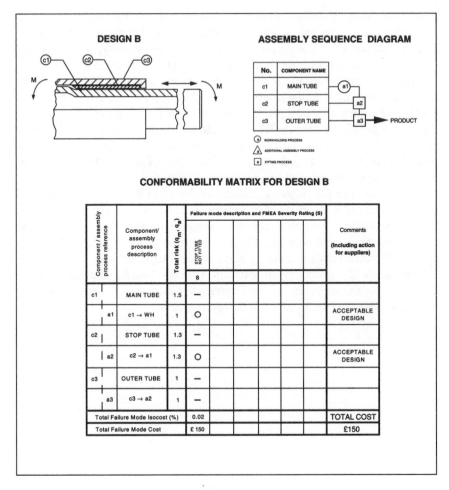

Fig. 7.11 Telescopic lever assembly – design B

inner diameter of the main tube (c1), the outside diameter of the reinforcement tube (c2) and the application of the adhesive (a3). Commercial
confidence precludes the inclusion of component drawings and detailed
analysis.

The conclusion from the analysis is that the assembly should be
redesigned. This is further justified by calculating the potential costs of
quality for the assembly. If this design of telescopic lever assembly frac-
tured in service, user injury, high losses, including legal costs, could be
incurred. A cost of quality of £274 000 was calculated from the analysis,
which is far too high, representing as it does more than 36% of annual rev-
enue from the product. This figure was calculated by summing the isocosts
for each characteristic/assembly process whose variability risks potentially

contribute to each failure mode, and then multiplying the total failure mode isocost (%) by the product cost and the number of items produced. The calculation of the costs for the second failure mode type (reinforcing tube moves out of position) on design A, is shown in detail in Figure 7.10.

Failure of this design in service resulted in user injury and high losses of the order of those calculated above, including legal costs, are incurred. A number of alternative designs are possible, and one that does not involve the above problems (design B) is included with its conformability matrix in Figure 7.11 (commercial confidence precludes the inclusion of component drawings and detailed analysis). The costs of quality for this design were subsequently reduced to a negligible amount.

The above example suggests that the analysis can be used to support the identification of conformance problems before production commences. Given the huge losses that can be associated with safety critical conformability problems, considerations of the type proposed must be on the agenda of concurrent engineering and product introduction.

Where in the product introduction process should the analysis be applied? The timing illustrated in Figure 6.9, Chapter 6, represents what is believed to be the best practice in this connection.

8 The costs of quality

Every time we fail to meet the needs of our internal or external customers it costs money to replace, rectify, pay warranties or recall products. As stated earlier, it is not uncommon for these costs to represent some 20% of turnover. This is entirely lost profit, hence the important principle of getting things 'right first time'.[64-66]

Not until recently have companies really attempted to determine the true cost of quality, mainly due to the lack of effective cost-control systems. The main problems of quality can be summarised as follows:

- most office and manufacturing processes produce defect levels around 20%;

- twenty per cent is accepted as the norm;

- no organised effort is made to change the situation until a 'real' quality problem is found;

- all efforts are then expended to solve the problem and return to the previously acceptable level.

8.1 Elements of the cost of quality

The overall cost of quality can be divided into the following four categories:

1 Prevention costs –
These are the expected costs of getting things right first time, e.g. the costs of quality in design, quality planning, quality assurance, quality training and quality improvement projects. Prevention costs are referred to as POC (Price Of Conformance), i.e. all those activities that ensure all subsequent actions are right first time.

2 Appraisal costs –
   These are the costs such as inspection and checking of goods and
   materials on arrival. Whilst an element of inspection is necessary and
   justified, it should be kept to a minimum as it does not add any value
   to the project.

3 Failure costs –
   (a)   Internal failure costs –
         These are essentially the costs of failures identified and recti-
         fied before the final product gets to the external customer.
   (b)   External failure costs –
         These include the return of defective goods, sending engi-
         neers to the customer to rectify deficiencies, warranties,
         product liability claims, etc.
   Both (a) and (b) represent the cost of nonconformance to requirements and are referred to as PONC (Price of NonConformance), i.e.
   costs that arise as a result of processes that are not right first time.
   Examples of possible POCs and PONCs in business are given in
   Appendix 2.

4 Lost opportunities –
   This category of quality is impossible to quantify accurately but basi-
   cally refers to the rejection of a company product due to a history of
   poor quality and service and where the company is not invited to bid
   for future contracts.

Therefore, any effective costing system can only determine the costs
incurred under items 1–3.

8.2   Quality cost models

Many organisations fail to appreciate the scale of their quality failures and
employ financial systems that fail to quantify and record the true costs of
quality; in many cases the failures are costs that are logged as 'overheads'.
Quality failure costs represent a direct loss of profit!

Organisations may have financial systems to recognise scrap, inspection,
repair and test; these only represent the 'tip of the iceberg', as illustrated
in Figure 8.1.

The gathering of all this data should not deteriorate into a statistical
exercise. The costs should be analysed as they form an invaluable tool for
highlighting areas of poor efficiency, making everyone aware of the problems. They should lead to strategy and plans for recovery, record the pace
of achievement, and form a major part of business planning (see Figure
8.2).

Figure 8.3 illustrates how a company used the costs of quality (albeit
only as good as its financial systems) as part of its strategy for improvement. The measured costs indicated 55% internal failures, 28% appraisal

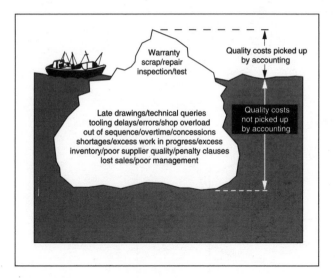

Warranty
scrap/repair
inspection/test

Quality costs picked up
by accounting

Late drawings/technical queries
tooling delays/errors/shop overload
out of sequence/overtime/concessions
shortages/excess work in progress/excess
inventory/poor supplier quality/penalty clauses
lost sales/poor management

Quality costs
not picked up
by accounting

Fig. 8.1 Hidden cost of poor quality

i.e. inspection and test, and 17% prevention. In its five-year plan, the company set an objective to reduce failures by 50%, reduce appraisal and slightly increase prevention to significantly reduce the costs of quality overall.

A company should minimise the cost of nonconformance and minimise appraisal costs, but be prepared to spend adequately on prevention. Some

- Track overall quality performance

- Compare cost results to goals

- Trend analyses, categories and accounts

- Investigate reasons for unusual costs

- Prioritise problems/opportunities

- Initiate programmes and reduction of costs

- Develop goals as a part of business planning

Fig. 8.2 Uses of quality costs

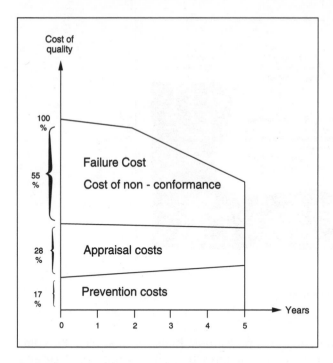

Fig. 8.3 Typical relationship between the various costs of quality

quality gurus strive for zero defects. Whilst this is obviously the ultimate goal, the costs can become prohibitive. It is possible to determine the optimum (from a cost point of view). This value is not constant across a company, product or service. For instance, a machine shop may be prepared to accept a 1% scrap rate, but it is doubtful that the public would accept that failure rate from a finished aircraft! Each must set its own objectives.

The optimum defect level will vary according to the application – the more severe the consequences of failure the higher the quality performance needs to be. A simple quality cost model that a business can develop to define an optimum between quality and cost is illustrated in Figure 8.4.

Economic quality/cost models can help a business to understand the influence of defect level on cost. The models tend to indicate general trends in quality cost and many have been produced over a number of years. They include notional diagrams indicating the presence of a minimum total quality cost against some general defect level, and models which suggest how the quality-cost elements – prevention, appraisal and failure – change with increasing quality awareness and improvement. A classification of models has been proposed by Plunkett and Dale[67] and the reader is also directed to BS 6143: Part 2 (1990)[68] and elsewhere.[69,70]

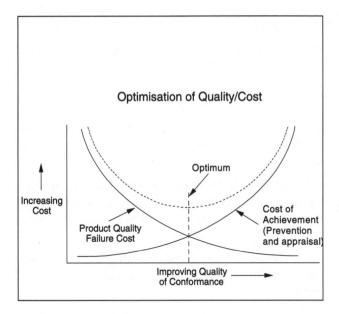

Fig. 8.4 Quality/cost optimisation

8.3   Industrial experiences 1

As stated elsewhere, typically a company may buy in perhaps two thirds of the cost of their final product in terms of materials, and subcontract the manufacture of proprietary items such as pumps, valves, electronic units, etc. Should these items fail on final assembly or test, the total cost of the quality failure could include:

- delay in final delivery incurring penalty costs to the external customer or interest on the capital involved with no value added;

- cost of removing the defective component and refitting a replacement unit (if available);

- inspection costs and the preparation of defect reports;

- re-packaging and cost of despatch to the manufacturer;

- administration costs incurred by the procurement and financial departments;

- contentious debate with the manufacturer who may query liability for the defect.

As can be seen, the true costs of the failure could be many times the costs usually measured.

8.4   Industrial experiences 2

This illustrates the possibility for a typical machine shop to identify and record the true costs of a machining failure, examples of which are given below.

• The machine is stopped, hence loss of production.

• The defective component is examined to determine whether it should be repaired or scrapped. If it is repaired, the costs should include technical work, fitting and inspection.

• Further costs not generally identified include a possible build up of work in progress incurring costs without value added, costs of manufacturing a new component if the original was scrapped, out of sequence working or overtime to recover the programme, etc.

Time spent on developing thorough control systems will provide management with better tools to identify quality problems and better direct resources to resolve them.

9 A review of selected quality tools and techniques

In this chapter the following quality tools and techniques are reviewed:

- Failure Mode and Effect Analysis (FMEA);

- Statistical Process Control (SPC);

- process capability studies (C_p and C_{pk});

- Design Of Experiments (DOE);

- Quality Function Deployment (QFD);

- Poka Yoke.

9.1 Failure Mode and Effect Analysis (FMEA)

Failure mode and effect analysis is a systematic element-by-element assessment to highlight the effects of a component, product, process or system failure to meet all the requirements of a customer specification, including safety. This indicates by high point scores those elements of a component, product, process or system requiring priority action to reduce the likelihood of failure through redesign, concurrent engineering, safety back-ups, design reviews, etc. It can be carried out at the design stage using experience or judgement, or integrated with reliability data using knowledge of failure rates for existing components and products.

The following factors are assessed in an FMEA:

- Potential failure mode – how could the component, product, process or system element fail to meet each aspect of the specification?

- Potential effects of failure – what would be the consequences of the component, product, process or system element failure?

- Potential causes of failure – what would make the component, product, process or system fail in the way suggested by the potential failure model?

- Current controls – what is done at present to reduce the chance of this failure occurring?

- Occurrence (O) – the probability that a failure will take place, given that there is a fault.

- Severity (S) – the effect the failure has on the user/environment, if the failure takes place.

- Detectability (D) – the probability that the fault will go undetected before the failure takes place (D_1) multiplied by the probability that the failure will go undetected before having an effect (D_2).

The occurrence, severity and detectability ratings are assessed on a scale of 1–10 and are illustrated below in Figure 9.1.

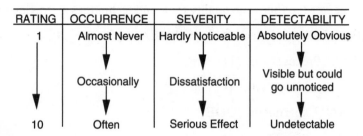

RATING	OCCURRENCE	SEVERITY	DETECTABILITY
1	Almost Never	Hardly Noticeable	Absolutely Obvious
	Occasionally	Dissatisfaction	Visible but could go unnoticed
10	Often	Serious Effect	Undetectable

Fig. 9.1 General ratings for FMEA occurrence, severity and detectability

For more comprehensive FMEA occurrence, severity and detectability ratings, see Tables 9.1, 9.2 and 9.3, respectively. Note, occurrence can be replaced by reliability data in the form of failure rates or Statistical Process Control (SPC) data in the form of a parts-per-million (ppm) failure rate once a process is in control.

The Risk Priority Number (RPN), is the occurrence (O), severity (S) and detectability (D) ratings multiplied together:

$$RPN = O \times S \times D$$

This number should be used as a guide to the most serious problems, with the highest numbers requiring the most urgent action.

In the case of design FMEAs, the RPN score can:

- highlight the need for design improvement;

- highlight the needs for SPC, 100% inspection, no inspection, etc.;

- highlight elements for full Taguchi treatment;

- highlight a priority order for focusing limited resources;

Table 9.1 FMEA definitions for occurrence rating

Occurrence (O)		
Rating	Guide	Possible failures (ppm)
1	Remote possibility of failure occurring	0.1
2	Low possibility of occurrence	0.5
3	Low possibility of occurrence	2
4	Moderate possibility of occurrence	10
5	Moderate possibility of occurrence	50
6	Significant number of failures possible	200
7	High possibility of occurrence	1000
8	High possibility of occurrence	5000
9	Very high possibility of occurrence	20 000
10	Almost certain that many failures will occur	100 000

Table 9.2 FMEA definitions for severity rating

Severity (S)	
Rating	Guide to effect on user
1*	No effect or minimal effect on customer
2*	Minor annoyance to customer
3*	Annoyance to customer but no loss of major function
4*	Possible return to manufacturer
5*	Definite return to manufacturer
6	Failure leading to violation of statutory requirement
7	Failure leading to injury or a more safety critical related problem with secondary back-up
8	Safety problem – degradation of function with possible severe injury
9	Complete failure with probable severe injury and/or loss of life
10	Catastrophic failure with high probability of loss of life

*These failures are non-safety critical

Table 9.3 FMEA definitions for detectability rating

Detectability (D)	
Rating	Guide
1	*Always obvious – foolproof
2*	Obvious to human senses
3*	Inspection effort required
4*	Careful inspection by human senses
5*	Very careful inspection by human senses
6	Simple aids and/or disassembly required
7	Inspection aids and/or disassembly required
8	Complex inspection and/or disassembly required
9	Very high possibility of non-detection
10	Undetectable

*Failure detection by human senses

- identify parts which have redundant function;
- prioritise suppliers listing as targets for supplier development;
- provide a basis for measures of performance.

9.1.1   Case study – design FMEA of a bicycle rear brake lever assembly

The design FMEA process is shown for a case study in Figure 9.2. It highlights the areas that need special attention, reflected in the highest risk priority numbers, when designing a bicycle rear brake assembly.

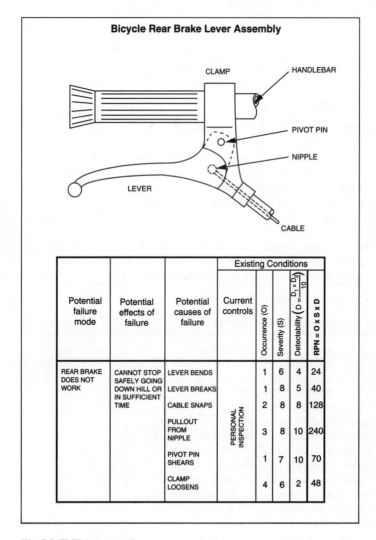

Fig. 9.2 FMEA case study

The analysis shows that design effort should be focused on the flexible element in the assembly, i.e. the brake cable. This perhaps supports the personal experiences of the reader, but the FMEA has shown this in a structured and rigorous appraisal of the concept design.

More information on FMEA and its applications is given elsewhere.[71–73]

9.2   Statistical Process Control (SPC)

By measuring small samples of components taken successively from a manufacturing process, it is possible to predict whether the process is in control or not without having to measure every component. There are two main types of statistical process control technique that are used for this purpose:

- mean control chart ($\bar{x}$) – see Figure 9.3;

- range control chart (R) – see Figure 9.4.

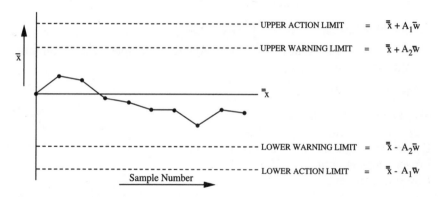

Fig. 9.3 Mean control chart

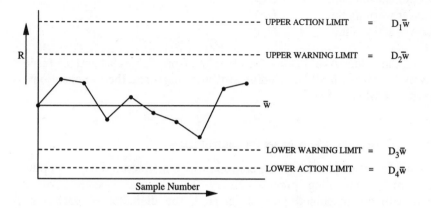

Fig. 9.4 Range control chart

Mean control charts plot the mean or average value ($\bar{x}$) of the sample data to detect drift in the process. Range control charts are used to detect process variability by plotting the sample range (R).

Warning and action limits on the control charts are initially set up from statistical data taken when the process is known to be running satisfactorily, i.e. in control. When these control limits have been set, the subsequent sample data, $\bar{x}$ and R, are plotted on the charts to observe the process over the next timescale. If during the process the value of $\bar{x}$ or R falls between the warning or action limits, then the operator knows the process is starting to go out of control and action needs to be taken to prevent nonconforming components from being produced.

To set up the control limits we need to calculate

Sample mean $\qquad \bar{x} = \dfrac{\Sigma fx}{n}$

where f = frequency, x = component variable, n = sample population

Grand mean $\qquad \bar{\bar{x}} = \dfrac{\Sigma \bar{x}}{N}$

where N = number of samples

Sample range $\qquad R = x_{\text{max}} - x_{\text{min}}$

where x_{max} = maximum value of x found in sample, x_{min} = minimum value of x found in sample

Mean of sample ranges $\quad \bar{w} = \dfrac{\Sigma R}{N}$

Mean control chart limits
Upper action limit $\quad = \quad \bar{\bar{x}} + A_1 \bar{w}$
Upper warning limit $\quad = \quad \bar{\bar{x}} + A_2 \bar{w}$
Lower warning limit $\quad = \quad \bar{\bar{x}} - A_2 \bar{w}$
Lower action limit $\quad = \quad \bar{\bar{x}} - A_1 \bar{w}$

Range control chart limits
Upper action limit $\quad = \quad D_1 \bar{w}$
Upper warning limit $\quad = \quad D_2 \bar{w}$
Lower warning limit $\quad = \quad D_3 \bar{w}$
Lower action limit $\quad = \quad D_4 \bar{w}$

Note, the values A and D are factors taken from Tables 9.4 and 9.5 respectively.[74,75] More on SPC, including attribute charts and their applications, is given elsewhere.[43,63,76,77]

9.2.1   Case study – how to set up control charts

Cylindrical steel pins are being continuously manufactured by a computer-controlled machining process. Every 30 minutes, a sample of five consecutively machined pins is selected, the diameter of each pin is measured and the sample mean and sample range are calculated. The 10

Table 9.4 Mean control chart factors

Sample size, n	Action factor, A_1	Warning factor, A_2
2	1.937	1.229
3	1.054	0.668
4	0.750	0.476
5	0.594	0.377
6	0.498	0.316
7	0.432	0.274
8	0.384	0.244
9	0.347	0.220
10	0.317	0.202

Table 9.5 Range control chart factors

Sample size, n	Upper action factor, D_1	Upper warning factor, D_2	Lower warning factor, D_3	Lower action factor, D_4
2	4.12	2.87	0.04	0.00
3	2.98	2.17	0.18	0.04
4	2.57	1.93	0.29	0.10
5	2.34	1.81	0.37	0.16
6	2.21	1.72	0.42	0.21
7	2.11	1.66	0.46	0.26
8	2.04	1.62	0.50	0.29
9	1.99	1.58	0.52	0.32
10	1.93	1.56	0.54	0.35

results over 25 samples are given in Table 9.6. Using the first ten samples, calculate the control limits for the mean and range charts assuming the process is in control and running normally.

Plot the values of the next 15 samples on (i) mean and (ii) range control charts using the limits calculated above. Comment on the behaviour of the process and any action required.

Calculations using the first ten samples to set control limits give

Grand mean $\qquad \bar{\bar{x}} = \dfrac{\Sigma \bar{x}}{N} = \dfrac{290}{10} = 29.00$ mm

Mean of sample ranges $\qquad \bar{w} = \dfrac{\Sigma R}{N} = \dfrac{0.6}{10} = 0.06$ mm

From Table 9.4, for sample size = 5, the 'A' factors give:

Mean control chart limits
Upper action limit $\ = \bar{\bar{x}} + A_1\bar{w} = 29 + (0.594 \times 0.06) = 29.036$ mm
Upper warning limit $= \bar{\bar{x}} + A_2\bar{w} = 29 + (0.377 \times 0.06) = 29.023$ mm
Lower warning limit $= \bar{\bar{x}} - A_2\bar{w} = 29 - (0.377 \times 0.06) = 28.977$ mm
Lower action limit $\ = \bar{\bar{x}} - A_1\bar{w} = 29 - (0.594 \times 0.06) = 28.964$ mm

Table 9.6 Sample data for pins

Sample number	Mean diameter (mm)	Range of diameters (mm)
1	29.01	0.07
2	29.01	0.07
3	28.99	0.06
4	29.00	0.04
5	29.00	0.05
6	29.02	0.05
7	28.98	0.06
8	28.99	0.07
9	29.00	0.07
10	29.00	0.06
11	29.00	0.06
12	28.99	0.05
13	28.99	0.04
14	29.01	0.06
15	29.01	0.06
16	29.02	0.06
17	29.01	0.08
18	29.01	0.09
19	29.02	0.08
20	29.02	0.10
21	29.02	0.11
22	29.03	0.12
23	29.02	0.12
24	29.03	0.12
25	29.03	0.15

Samples 1–10 are marked "Process in control".

From Table 9.5, for sample size = 5, the 'D' factors give:

Range control chart limits
Upper action limit $= D_1\bar{w} = 2.34 \times 0.06 = 0.140$ mm
Upper warning limit $= D_2\bar{w} = 1.81 \times 0.06 = 0.109$ mm
Lower warning limit $= D_3\bar{w} = 0.37 \times 0.06 = 0.022$ mm
Lower action limit $= D_4\bar{w} = 0.16 \times 0.06 = 0.010$ mm

The mean and range charts are constructed using the above control limits. The sample data, $\bar{x}$ *and* R for samples 11–25 are then plotted on the mean and range control charts as shown in Figures 9.5 and 9.6, respectively.

The mean control chart indicates possible tool wear, but the range control chart indicates a gradual range increase. This could mean something working loose, built-up edge effects, material variability influencing surface finish, vibration effects, etc. Action will need to be taken to rectify the situation before nonconforming products are manufactured.

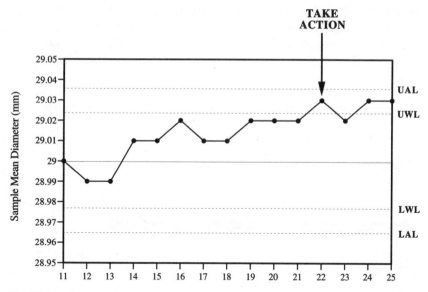

Fig. 9.5 Mean control chart – case study

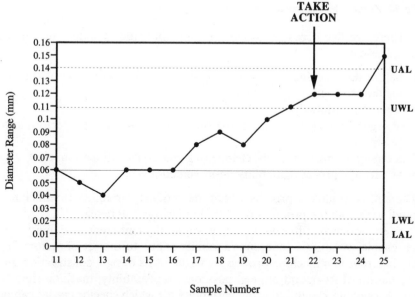

Fig. 9.6 Range control chart – case study

9.3 Process capability studies (C_p and C_{pk})

A capability study is a statistical tool which measures the variations within a process. Samples of the product are taken and measured and the variation is compared with the tolerance. This comparison is used to establish how 'capable' the processes are in producing the product.

Process capability is attributable to a combination of the variables in all of the inputs. Machine capability is calculated when the rest of the inputs are fixed. This means that the process capability is not the same as machine capability. A capability study can be carried out on any of the inputs by fixing all the others.

All processes can be described by Figure 9.7 where the distribution curve for a process shows the variability due to its particular elements.

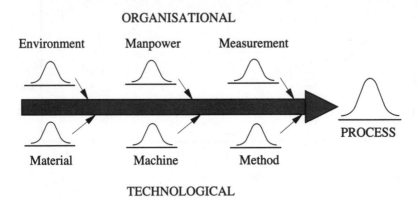

Fig. 9.7 Factors affecting process capability

There are five occasions when the capability studies should be carried out. These are:

1 before the machine/process is bought (to see if it is capable of producing the required components);
2 when it is installed;
3 at regular intervals to check that the process is giving the performance required;
4 if the operating conditions change (e.g. materials, lubrication, etc.);
5 as part of a process capability improvement.

The aim is to have a process where the product variability is sufficiently small so that all the products produced are within tolerance.

There are two different kinds of variability: inherent variability and assignable variability. Inherent variability is due to the set of factors that are inherent in a machine/process by virtue of its design, construction and the nature of its operation, e.g. positional repeatability, machine rigidity, etc. Assignable variability is the variability for which specific causes can be identified. It can be eliminated or minimised through techniques like process capability. For a process to be capable of producing components to specification, the sum of inherent and assignable variability must be less than the tolerance.

Capability is measured carrying out a capability study and calculating a capability index. There are two common process capability indices, C_p and C_{pk}.

9.3.1 Process Capability Index, C_p

The process capability index is a means of quantifying a process to produce components within the tolerances of the specification.

$$C_p = \frac{U - L}{6\sigma}$$

where, U = upper control limit, L = lower control limit and σ = standard
deviation of population (a measure of the dispersion or spread of the
population data). A value of $C_p = 1.33$ would indicate that the distribution of the products covers 75% of the tolerance. This would be
sufficient to assume that the process is capable of producing everything to
specification.

The machine/process capability index, C_p is interpreted as follows:

| less than 1.33 | → | process not capable |
| between 1.33 and 2.5 | → | process capable |

Where a process is producing a product with a capability index greater
than 2.5, it should be noted that the unnecessary precision may be
expensive.

Figure 9.8 shows process capability in terms of the tolerance of a component. The variability of the data does not always take the form of a normal distribution. There can be 'skewness' in the shape of the distribution
curve, this means the distribution is not symmetrical, leading to a lopsided
appearance. The 'skewness' can be accounted for by using C_{pk}.

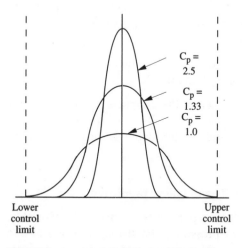

$C_p = 2.5$

$C_p = 1.33$

$C_p = 1.0$

Lower control limit

Upper control limit

Fig. 9.8 Process capability in terms of tolerance

9.3.2 Process Capability Index, C_{pk}

By calculating where the process is centred (the current mean value) and
taking this, rather than the target value, it is possible to account for the

skew of a distribution (see Figure 9.9) which would render C_p inaccurate. C_{pk} is calculated using the following equation:

$$C_{pk} = \frac{|\mu - L_n|}{3\sigma}$$

where, μ = mean value of population, L_n = nearest control limit and σ = standard deviation of population. By using the nearest control limit, the worst-case scenario is being used, ensuring that over-optimistic values of process capability are not employed.

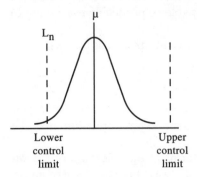

Fig. 9.9 Process capability and skew

C_{pk} is a much more valuable tool than C_p as it can be applied accurately to both skewed and normal distributions. As a large percentage of distributions are skewed, C_p is limited in its usefulness. If C_{pk} is applied to a non-skewed normal distribution, by the nature of its formula it reverts to C_p.

C_{pk} is interpreted in the same way as C_p:

less than 1.33   →   process is not capable
between 1.33 and 2.5   →   process is capable

Again, for $C_{pk} > 2.5$, it should be noted that the unnecessary precision may be expensive.

For more on process capability studies, the reader is directed to other references.[44,63,78]

9.3.3   Case study – process capability and failure prediction

The component shown in Figure 9.10 is part of an electric light assembly. The component is manufactured by pressing a flat strip at the rate of one million per annum. Using the statistical process control data for the component characteristic (A) shown in Figure 9.11, we can calculate the process capability index, C_{pk}. (Note: Figure 9.11 shows a skewed distribution, so C_{pk} is chosen.) Using the graph on Figure 9.12 (which shows the relationship between C_{pk} and parts-per-million failure rate), the likely annual failure rate of the product can be calculated.

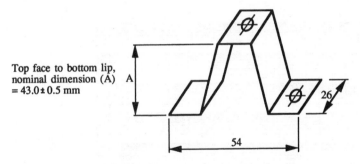

Top face to bottom lip,
nominal dimension (A)
= 43.0 ± 0.5 mm

Fig. 9.10 Component design

The solution is as follows:

$$\mu = \frac{\Sigma fx}{N} \quad = \quad \frac{2 \times 42.5 + 4 \times 42.6 + 10 \times 42.7 + 4 \times 42.8 + 8 \times 42.9 + 8 \times 43.0 + 2 \times 43.1 + 2 \times 43.2}{40}$$

$$= \quad 42.84 \text{ mm}$$

$$\sigma = \sqrt{\frac{\Sigma f(x - \mu)^2}{N}}$$

$$= \sqrt{\frac{\begin{array}{l}2 \times (42.5 - 42.84)^2 + 4 \times (42.6 - 42.84)^2 + 10 \times (42.7 - 42.84)^2 + 4 \times (42.8 - 42.84)^2 \\ + 8 \times (42.9 - 42.84)^2 + 8 \times (43.0 - 42.84)^2 + 2 \times (43.1 - 42.84)^2 + 2 \times (43.2 - 42.84)^2\end{array}}{40}} = 0.18$$

$$C_{pk} = \frac{|\mu - L_n|}{3\sigma} = \frac{42.84 - 42.5}{3 \times 0.18} = 0.63$$

where f = frequency, x = component variable, N = population, σ = standard deviation of population, μ = mean value of population, C_{pk} = process capability index and L_n = nearest control limit.

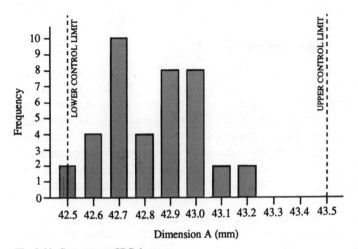

Fig. 9.11 Component SPC data

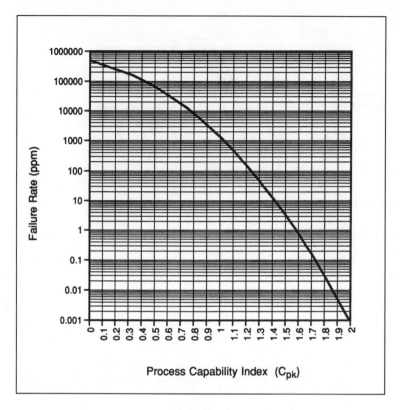

Fig. 9.12 Relationship between C_{pk} and failure rate (ppm)

At $C_{pk} = 0.63$, from Figure 9.12 the likely annual failure rate would be approximately 30 000 ppm. Of course, action would be taken to prevent further defects from being produced and to avoid this failure rate in the future.

9.4   Design of Experiments (DOE)

DOE encompasses a range of techniques used off-line to enable a business to understand the effects of important variables in product and process development. It is normally used when investigating a situation where there are several variables, one or more of which may result in a problem either singly or in combination. It complements quality control techniques such as SPC, homing in on those areas where control is necessary. Its application logically follows on from DFQ and FMEA techniques where particular design or process characteristics are identified as requiring further investigation.

The variables or factors are identified using engineering judgement or brainstorming. Having identified the relevant factors, experiments are conducted by exploring the effect of their 'levels'. The factors are measurable units, such as pressure, viscosity or processing time, that have an influence on the quality level of a product or process. The levels are the values or settings of the variables.

The complete set of possible combinations of factors and levels in an experiment is the 'full factorial' and is often too large to be considered. If there are 13 variables each with three settings or levels, then over 1.5 million trials would be necessary to explore all combinations. DOE aims to determine the subset of the 'full factorial' that should be selected for study. The traditional approach is to explore one factor at a time, while all the others are maintained constant, but obviously this can give an unsatisfactory conclusion since the effect of interactions between the factors is not considered. Alternatively an 'orthogonal array' coupled with analysis-of-variance techniques can be used to focus in on the key combinations.

9.4.1 An application methodology

The systematic application of DOE should be based on three distinct phases, described below.

9.4.1.1 Phase 1 – preparation

This is often referred to as the pre-experimental stage. Experiments can take considerable time and resources, and good preparation is all important. DFQ and FMEA results are important inputs to this phase of the methodology, providing focus and priority selection filtering. A summary of the steps that should be considered is given below.

- Define the problem to be solved.

- Agree the objectives and prepare a project plan.

- Examine and understand the situation. Obtain and study all available data related to the problem.

- Define what needs to be measured to satisfy the project objectives.

- Identify the factors to be controlled during the experiment and select the levels to be considered. These should be representative of normal operating range and sufficiently spaced to spot changes.

- Establish an effective measuring system. Understand its variance and the likely effects of this apparent variation in output.

9.4.1.2 Phase 2 – experimentation and analysis

Carrying out the planned trials and analytical work will include the steps below.

- Select the techniques to be used. Decisions on the number of trials are invariably coloured by cost, but always check that they are balanced, otherwise it may be false economy.

- Plan the trials. Consecutive trials should not affect each other.

- Conduct the trials as planned. Results should be carefully recorded in a table or machine along with any observations that may be regarded as potential errors.

- Analyse and report the results. The analysis of results should be transparent. Simple approaches should be used where possible to build confidence and provide clarity.

9.4.1.3   Phase 3 – implementation

This phase is often called the post-experimental stage. It involves acting on the results and communicating the lessons learnt. Some points on the process are given below.

- Apply the findings to resolve the problem.

- Adopt a procedure for measuring and monitoring results such as SPC to detect future changes that could influence quality.

- Communicate lessons learnt through the business and include them in training programmes.

Other reviews of DOE[24,79] have been helpful in the preparation of this overview. More detailed information is given elsewhere.[80–82]

We shall now go on to consider a simple case study to explain the technique in more detail.

9.4.2   Case study – maximisation of flow through a filter

Consider the example of fluid flow through a filter, where the objective is to maximise flow rate. Assume that the relevant factors are filter, fluid viscosity and temperature, and that each can have two settings, coarse/fine, high/low and high/low, respectively. Figure 9.13(a) shows a full factorial experiment where all combinations of factors and levels are explored based on eight trials. As touched on previously, this approach can be costly. In Figure 9.13(b) the common method of varying a single factor is illustrated. This approach does not properly hit the interaction effects. In a more balanced experiment, the different levels of each factor would occur on the same number of occasions. Figure 9.13(c) illustrates an orthogonal array involving four trials. Note that between any two columns each combination level occurs equally often. Using this approach the effect of the different factors can be separated out and their effects estimated.

In many practical cases, useful experiments can be carried out using only two levels. Often this can be employed where a single change in level will illustrate whether a factor is likely to be significant or not. However, where a response is non-linear, three or more levels may be needed.

When the experiment is complete it is time to investigate the significance of each factor. This can be done with the aid of straightforward

a)

Factor Trial No.	A Filter density	B Fluid viscosity	C System temperature	Result*
1	coarse	low	low	8
2	coarse	low	high	10
3	coarse	high	low	4
4	coarse	high	high	6
5	fine	low	low	4
6	fine	low	high	6
7	fine	high	low	2
8	fine	high	high	4

Full factorial exploring all combinations of variables and their levels. This approach can prove to be costly and time consuming.

b)

1	coarse	high	high	6
2	fine	low	high	6
3	fine	high	low	2

Varying a single factor only can miss important interactions.

c)

1	coarse	low	low	8
2	coarse	high	high	6
3	fine	low	high	6
4	fine	high	low	2

Orthogonal Array involving 4 trials. Here the effect of different factors can be separated out and their effects estimated.

Fig. 9.13 Experimental variations for fluid flow through a filter

statistical or graphical analysis-of-variance techniques. The graphical approach is particularly simple and effective and the approach is outlined in Figure 9.14. In this figure, values from Figure 9.13(c) are used to generate plots of factors and levels. The plots are employed to study the effects of both individual results and combinations. In the case of individual results (Figure 9.14(a)), the gradient of the line drawn between average values of the results at the levels under consideration is a measure of the significance of the factor. A high gradient indicates a high significance, while a horizontal line infers no significance at all. Analysis of the effects

a) Individual results

The gradient of the graph indicates the significance of the factor.

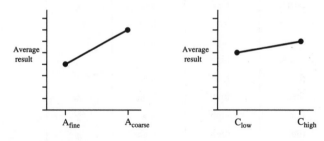

The high differential means
highly significant.

A low differential means low
significance. A zero differential
means not significant at all.

b) Combination of results

The relationship of the lines shows the significance of the combination.

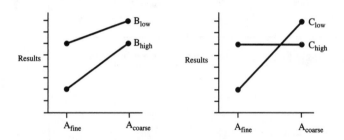

Significant combinations of factors will be indicated when the
lines show a large difference in gradient. Strong significance is
indicated where the lines cross. Parallel lines indicate little
significance.

Fig. 9.14 Graphical analysis of experimental results

of combinations is based on lines constructed as shown in Figure 9.14(b).
The relationship between the lines indicates the significance of the results.
Significant combinations are inferred when the lines show a large difference in gradient. Strong significance is indicated when lines cross. Parallel
lines suggest no significance.

9.5 Quality Function Deployment (QFD)

There is currently much interest in QFD and it is finding application in
some major businesses round the world. In QFD customer requirements
or the 'voice of the customer' as they are sometimes called, are cascaded

down through the product development process in four separate phases (see Figure 9.15). In phase 1 of QFD, customer requirements or attributes (inputs) form the rows of a matrix structure where the columns are represented by product design features and functions. Conventions are then used to define the relationship between the matrix elements together with the customer priorities. A popular convention is shown in Figure 9.16, in this case for a baggage conveyor. This yields the all-important quantification between customer requirements and product design issues, albeit a subjective quantification.

The need is then to transfer the important product features and functions to the rows of the next matrix where the columns are the component characteristics (see Figure 9.15). In this way, customer requirements are cascaded down through the product introduction process, keeping the effort focused on the important issues, and thus a direct link is formed between the 'voice of the customer' (design requirement inputs in phase 1) and the actual shop-floor operations/procedures (outputs from phase 4). QFD is best suited to teamwork, having valuable team-building qualities, and represents a significant contribution beyond the more traditional design methods.

A further development to QFD relates to the inclusion of a number of enhancements, namely the analysis of complex products on different levels, such as system, subsystem and component; a status evaluation method, which indicates whether a concept is static or dynamic; contextual analysis; bench-marking techniques such as parameter analysis can be used to plot two characteristics against one another; and the use of a generic structure of requirements, to help with the identification and compilation of the needs. Although QFD has already been applied with additional techniques, enhanced QFD (EQFD)[69] formalises the enhancement to form an integrated procedure to further increase the probability that the right product will be developed.

Despite the enhancements mentioned, the main problems still faced in the application of QFD include the following: properly defining the appropriate matrix columns and rows for the cascade, although this is mainly down to training and experience; and the subjectivity of the importance rating procedure.

There are many texts covering the application of QFD. More detailed examples are given elsewhere.[83-6]

9.6 Poka Yoke

The provision of effective quality assurance to prevent defective products being produced is essential and in order to provide the emphasis on quality it should be moved from inspecting to preventing defects from being dispatched to the customer.

In the drive towards zero defects, the use of Poka Yoke (literally, mistake proofing), or foolproofing, within all stages of the production process

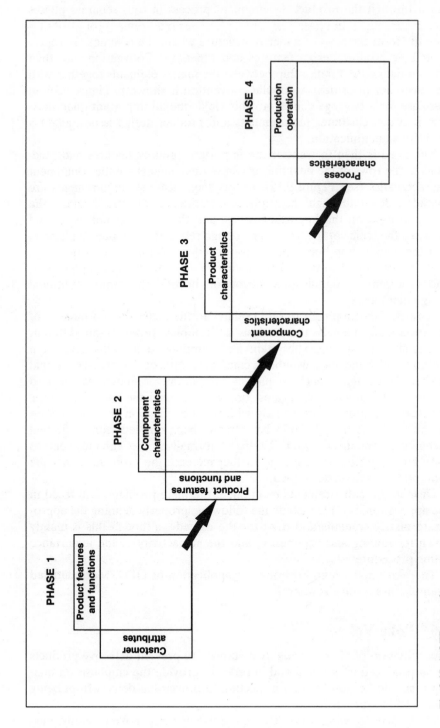

Fig. 9.15 A QFD cascade

Key:		Customer priority ratings	Product features and functions								
			Angle of conveyor to be minimum	Good grip belt to be used	Determine minimum speed of conveyor	Ensure belt is good fit and is aligned	Conveyor to plug into press socket	Non-fouling hood required on chute	130 mm diameter wheels to be used	Mount motor as low as possible	Use quick-release mechanisms
△ Weak relationship = 1											
○ Strong relationship = 3											
● Very strong relationship = 9											
Customer attributes	Baggage doesn't fall back down conveyor	4	○	●	○	△					
	Standard speed required on all conveyors	2			●						
	Baggage does not jam on conveyor	5	△	△	△	●			●		
	Conveyor to be moved easily	1					●	●	△	○	●
	Stop baggage bouncing off conveyor on to floor	3	○	○	○				●		
	Conveyor to be stable	4	○							●	
	POINTS RATING		38	50	44	49	9	9	73	39	9

Fig. 9.16 Relationships between the matrix elements for QFD

is essential and needs to be considered and implemented wherever possible. The key to Poka Yoke is the belief that quality inspections will never eliminate errors. The only way to eliminate errors is to stop making mistakes. Most operator mistakes are not the result of inattention or poor training, but because the system or process has been badly designed. If Poka Yoke techniques are applied to a manufacturing process or assembly operation, then the possibility of mistakes occurring is eliminated.

The core of Poka Yoke is the use of self-checks, successive checks and source inspections.

- Self-checks – the best person to detect mistakes is the operator carrying out the operation; this is referred to as a self-check.

- Successive checks – in an assembly line, the check should be carried out by the next operator; this is referred to as successive checking.

- Source inspections – the best time to detect a mistake is immediately after it has happened. Hence, the inspection should occur at the source of a mistake. The system should be designed to highlight or prevent mistakes. Mistakes will not turn into defects if operator errors are discovered and eliminated (or prevented) as they occur.

The range of possible Poka Yoke devices that could be applied is vast. Three general categories have been identified (see Figure 9.17).

- Contact type – the use of shape, dimensions or other physical properties of products to detect the contact or non-contact of a particular feature and hence prevent the manufacture of defects (e.g. asymmetrical holes through which the component must pass).

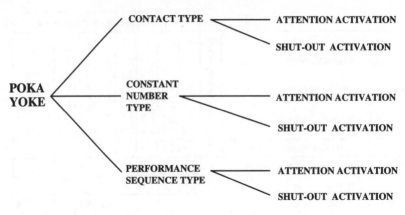

Fig. 9.17 Poka Yoke and methods of activation

- Constant number type – detects errors if a fixed number of movements have not been made (e.g. all parts have not been provided in a kit).

- Performance sequence type – detects errors if the fixed steps in a sequence have not been performed, or alternatively, prevents incorrect operations from being performed, thus eliminating any defects.

There are two recognised ways in which a Poka Yoke may be activated.

- Shut-out type – prevents incorrect action from taking place.

- Attention type – brings attention to an incorrect action but does not prevent its execution.

As the shut-out type of activation halts processing even if the operator is not paying attention, it is to be preferred (where possible) to attention type activation where production continues if the operator does not notice the warning.

Poka Yoke, while primarily intended as a means of ensuring that all stages in the production process are carried out correctly, in terms of product physical properties, numbers of movements and/or components, and errors in the steps involved in a sequence of operations, there are clear implications for the designer. For example, consideration should be given to component design such that they cannot be assembled in a way that would lead to the production of a defective product. The input/output analysis for Poka Yoke is shown in Figure 9.18.

More information on Poka Yoke and its applications can be found elsewhere.[78,87]

9.6.1 Case study 1 – welding jig

A company manufactured a plate component for braking systems. The plate was positioned onto a jig with the welding projections on top and then welded. A visual check was made prior to welding to ensure that the

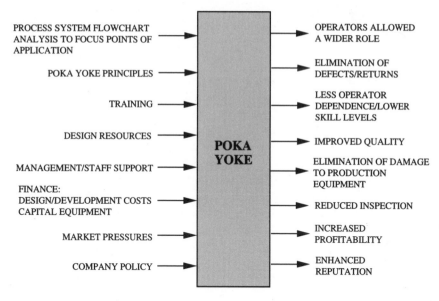

Fig. 9.18 Input/output analysis or Poka Yoke

plate was placed in the jig with the correct side up. However, 3–4% of plates were still welded upside down. As part of a Poka Yoke system, lugs were designed into the plate so that it would only fit into the jig when it was the right way up. This eliminated the possibility of upside-down errors occurring.

9.6.2   Case study 2 – non-return valve

An integral and safety critical component of an aircraft's landing gear system, a non-return valve, is made by a hydraulics company. The designers realised that the valve could be assembled the wrong way round, i.e. against the direction of flow of hydraulic oil, which would create a safety problem. An arrow was already marked on the valve showing the direction of flow, but this was felt inadequate under the circumstances. To overcome this problem, Poka Yoke principles were applied and the valve was redesigned with two different sizes of thread at the input and output ports to provide installation personnel with a further indication of the direction in which the valve operates in service.

10 Effecting a quality change

Any attempt to produce a quality change within an organisation will only be successful if:

- the need is fully recognised by the Managing Director (MD);

- the process of change is led by the MD;

- the MD appoints a quality director/executive. (The role of the quality director has been discussed in Chapter 2.)

10.1 Strategy for change

Comprehensive communication to staff is vital for success. Staff must be fully briefed at each stage of the activities in order that they fully support the changes and understand how they will concern everyone and when implemented should ensure a sustainable and prosperous future for all.

A programme of activities to effect the change should adopt the following strategy.

10.1.1 Review

A complete review of the quality in the organisation should be undertaken. The review should be co-ordinated by a senior manager who will give guidelines, help and requirements; however, it is essential that the departments themselves review their activities. It is suggested that the departments attempt to answer the following questions:

- How do you define quality?

- Do you consider that your department is achieving quality?

- What is the cost of achieving the quality?

10.1.2　Plan

From the review findings the senior management will prepare and approve a quality improvement plan including cost, major areas to be targeted initially, training plans, objectives to be achieved, etc. Do not be too ambitious at the start of the change. Select and concentrate on key factors rather than attempt too many changes at once.

10.1.3　Training

In every case it is anticipated that a programme of quality awareness and training will be necessary. This training must start with the directors and will cascade down the organisation to senior managers, first-line managers and operators. The training must not descend the pyramid until each level is fully conversant with it and committed. The training can be completed more effectively with the aid of facilitators who 'double-up' as quality leaders in their own departments. (See training pyramid in Figure 10.1.) It is important that the first-line supervision is totally committed and fully conversant with the modern quality tools and techniques in order to facilitate the change throughout the workforce (office or shop floor).

A typical training programme should start with the purpose and need for the training and should include these features:

- quality awareness;

- quality and customer satisfaction;

- brief history;

- customer/supplier relationship;

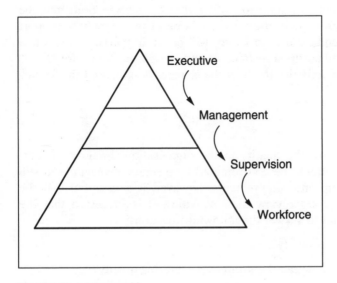

Fig. 10.1 Training pyramid

- quality costs;
- quality management systems and organisation;
- quality tools and techniques, including DFQ;
- the need for staff training;
- setting up quality initiatives and action teams;
- 'right-first-time' operation;
- supplier selection and performance.

The above concerns a vast range of subjects. Obviously it cannot all be communicated at once and it is suggested that the content be subtly tailored to different types of staff. It is recommended that each training session should not last for more than one day and that the programme should be spread over at least three to four weeks.

10.1.4   Implementation

Once the training is underway implementation will begin i.e. better identification of problems, solution of the problems and a programme of improvement/rectification. Metrics will have to be prepared to 'benchmark' the original poor quality and hence measure the improvements. 'Cost-of-quality' budgets should be introduced and included in the company business and profit plans.

10.2   The process of change

The process of change is not a 'one-off act' – it is a continuous process. The process can be divided into four phases.

1 Awareness and focus on the customer, i.e. training, identifying the internal customer, commitment and policy.
2 Assessment of the problem, human and organisational capabilities; do not underestimate what is invariably a huge change to the culture of the organisation.
3 Setting up the machinery of sustainable improvement; quality assurance, certification, processes and quality improvement plans.
4 Evaluation of the change, measurements of cost and achievement; determination of future improvements to ensure continuous improvement.

It is recommended that a company regards the four phases as a continuous process and repeats them every two to three years. How to effect the change is summarised in Figure 10.2.

The majority of companies embarking on improved quality programmes will experience huge cultural or behavioural changes. People quite

```
┌─────────────────────────────────────────────────┐
│                                                   │
│    1    Review                                    │
│                                                   │
│                                                   │
│    2    Prepare plan and present                  │
│                                                   │
│                                                   │
│    3    Training and awareness                    │
│                                                   │
│                                                   │
│    4    Implement and measure results             │
│                                                   │
│                                                   │
│    5    Review annually                           │
│                                                   │
└─────────────────────────────────────────────────┘
```

Fig. 10.2 How to effect a change

knowingly pass defective work to their customers without accepting any responsibility for the quality of their work. Therefore, the elements of a quality cultural change should include the following ethics:

- quality not cost as first priority;

- quality as everyones' responsibility;

- quality in every function of the organisation;

- quality changes management led;

- quality a process of continuous improvement;

- quality prevention and not correction.

10.3   A quality improvement programme

The implementation of a total quality culture or a process to improve company quality must be carefully introduced, otherwise it will be treated as 'yet another new management initiative'. Many companies have failed by trying to introduce the culture everywhere at the same time. It is recommended that key areas such as engineering, procurement and manufacturing should be selected for initial action, followed later by other departments.

When the first round of staff training is complete, department Quality Improvement Plans (QIPs) will be defined and Quality Action Teams (QATs) established to deal with the improvement proposals identified by the newly trained workforce. Baseline standards will be established and agreed improvement objectives will be placed on all departments. There may typically be a reduction in measured failure costs of 50% in three years. A departmental QIP is shown in Figure 10.3.

- The department quality improvement plan has to be an integral part of the business improvement plan.

- For successful implementation, quality improvement must become part of the day to day activity and not a series of individual projects.

- Typically a department quality improvement plan will include:

  - A clear statement of departmental quality policy

  - Identification of customers and suppliers

  - Analysis of current performance

  - Definition of problem areas

  - Development of corrective action teams which will produce:

    - detailed plans including implementation timing, resource requirement and accountability

    - a procedure for monitoring and auditing results

Fig. 10.3 Departmental quality improvement plan

The operation of the QATs is a departmental responsibility with the quality assurance department guiding and co-ordinating at the company level.

As stated previously, the quality improvement is not a 'one-off' but a continuous process which must be reviewed annually to determine progress against objectives and benefits gained against spend, to identify priority areas and most of all to keep the momentum going. Many companies will put much effort into rectifying short-term sporadic quality failures rather than use their assets to achieve permanent solutions (see Figure 10.4). They simply react to the problems and make little effort to prevent them. Very often a 'trouble shooter' is appointed and given large resources to effect a quick resolution. The 'trouble shooters' are very visible to top management and unfortunately gain greater kudos than those staff who are trying to prevent the occurrence of failures.

To summarise, the goal should be customer satisfaction at competitive cost (see Figure 10.5).

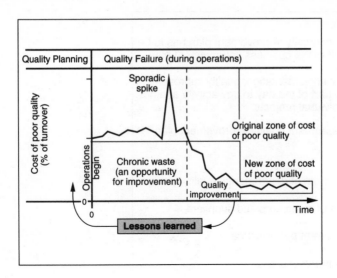

Fig. 10.4 Short-term versus long-term solutions

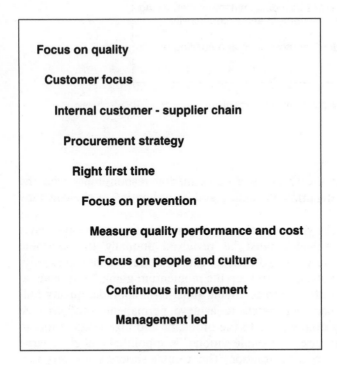

Fig. 10.5 Achieving customer satisfaction at competitive cost

11 Future developments

Whilst many important aspects of quality engineering are now in general practice, there are many outstanding issues and areas for research and development in both business and the academic community.

11.1 Extension of total quality management

Total quality is now being accepted in the UK, e.g. BS 7850: Parts 1 & 2 of BSI Total Quality Management (TQM). While TQM is being adopted by many large organisations, it is essential that smaller businesses take these systems on board. In several cases small-to-medium enterprises (SMEs) are driven to TQM by their links as suppliers to larger organisations. SMEs supplying direct to the public or to external customers do not generally have access to or understand the benefits that can be obtained by the introduction of TQM. The problem is being addressed by a number of institutions and professional bodies who are setting up 'quality centres' in the regions and national programmes of awareness. Whilst such initiatives are helpful, the main issue is of management culture. Most SMEs face many short-term problems, mostly cash flow, and often fail to see the benefits of what appears to be a long-term goal. Perhaps a better approach to TQM is to make companies aware of the costs of poor quality and its effects on the profitability of the company. Work, therefore, needs to be focused on identifying the cost of poor quality (work done incorrectly, poor service, warranties, etc.) which are simply lost profit. Total quality is about customer satisfaction at a competitive cost.

11.2 Initiating joint projects with suppliers and customers

SMEs could do well to copy the practice of the world's leading businesses and make closer links with both customer and supplier. Larger businesses are developing 'favoured suppliers' who become part of the product team. As previously mentioned, up to 60% of a product may be 'bought in' and it is important that interaction takes place at every level to ensure supplier quality.

In a similar way, some large companies include the customer in the product development design team. The problem here can be identifying who is the real customer for the product or service: the buyer, the product engineer, or the general manager. In most organisations contact with the external customer is initially with marketing and after delivery with customer support. These two organisations must feed the customer's wishes and satisfy performance requirements.

11.3   Research directions

Many gaps remain in our knowledge and a number of areas can be identified where work is necessary. A few such topics are listed below:

- research in DFQ and its application in product development;

- TQM in all aspects of business, e.g. personnel, finance, commercial, etc., and not least, product development;

- software quality assurance;

- integration of quality with technological development and business goals;

- manufacturing and process development.

Quality is largely built in (or not!) early on in design and tends only to be recognised much later. Businesses acknowledge that to reach the required quality objectives, changes are required in production and after launch. Therefore, DFQ is a key target. More research is needed on strategies, principles and techniques for DFQ covering both design quality and quality of conformance. We need to understand how quality is created from the needs of the customer and where it lies in the product. How does DFQ link with other quality/concurrent engineering tools and techniques, and where does it fit in the product development cycle? In designing a product, companies tend to look towards technology for the solution instead of getting the basics correct. DFQ knowledge-based techniques must be developed to provide for attention to detail and structure throughout the product development process. The ultimate aim of such techniques is to establish quality-orientated product development that reduces failure costs, reinforces customer confidence and supports supplier relationships. For a reference point on research in DFQ the reader is referred to the work of Mørup.[57]

Much has been written about research in total quality management. A review of the research challenges and research policy going back several years can be found elsewhere.[88,89] We need to understand the impact of leadership and culture on the success of TQM projects. Documentation of experiences and attitudes to TQM across business sectors would be helpful. Also, we urgently need more on the exploration of TQM in the management of the product development process.

The authors' experience in software development and managing projects suggests that there are many benefits to be gained from the development and application of quality tools and techniques for software development. Software Quality Function Deployment (SQFD) for software quality assurance, and SPC, attribute charts and associated statistics for error testing in software quality control are two examples where more research would be useful. What has been said above about the documentation of TQM experiences would also be helpful here.

Skills and technological leadership are often seen as independent and ultimate business goals. Technological leadership and leadership in quality are not always synonymous. How should these be focused to achieve quality improvement? How should research and development in new technology be linked to product development? Should they be separate?

While in most manufacturing businesses the bulk of components are bought in, little effort appears to be aimed at understanding the development of improved supplier relationships. More research is needed here. Research should also be directed at integrated manufacturing and quality control. Advanced production machines or cell systems with facilities for SPC-based gauging or measuring of important characteristics, to check whether they conform to specification, would be highly beneficial.

References

1 Quality Assurance Programme 1992–1996 (DTI Translation 0219-92, London). Federal Ministry of Research and Technology, Germany.
2 Juran, J. (1993) Made in USA: A renaissance in quality, *Harvard Business Review*, July/August, pp. 42–50.
3 Bendell, A. *et al.* (1993) *Quality: Measuring and Monitoring.* Century Business, London, UK.
4 Gitlow, H. and Gitlow, S. (1987) *The Deming Guide to Quality and Competitive Position.* Prentice Hall, New York.
5 Juran, J. (1988) *Juran on Planning for Quality.* The Free Press, New York.
6 Ishikawa, K. (1985) *What is Total Quality Control? The Japanese Way.* Prentice Hall, London.
7 Crosby, P. (1989) *Let's Talk Quality.* McGraw Hill, New York.
8 Feigenbaum, A. (1991) *Total Quality Control*, 3rd Edition. McGraw Hill, Maidenhead.
9 Shingo, S. (1986) *Zero Quality Control: Source Inspection and the Poka Yoke System.* Productivity Press, Cambridge, MA.
10 Peace, G.S. (1993) *Taguchi Methods: A Hands-on Approach.* Addison Wesley, Reading, MA.
11 Bendell, A. (1991) *The Quality Gurus.* Department of Trade and Industry, London, UK.
12 Swift, K.G. and Allen, A.J. (1992) Design for quality experiences, in *Proceedings WDK International Workshop on Design for Quality*, Technical University of Denmark, Lyngby, Denmark.
13 Hutchins, D.H. (1985) *Quality Circles Handbook*, Pitman, London.
14 Forward, G. (1989) Group activities the Japanese perspective, *Professional Engng*, July/August, pp. 24–25.
15 Sherwood, K.F. *et al.* (1993) Quality circles and total quality: a case study, *Total Quality Management*, **4**, No. 2, pp. 151–158.
16 Upda, S.R. (Ed.) (1990) in *Proc. Quality Circles India*, McGraw Hill, New Delhi.
17 Cooper, R. and Jensen, D. (1989) New organisational structures for greater competitiveness—a total quality approach, *Automotive Engineer*, April/May, pp. 52–53.
18 Maznevski, M.L. (1994) Understanding our differences: performance in decision-making groups with diverse members, *Human Relations*, **47**, No. 5, pp. 531–52.
19 Daniels, J.M. and Dale, B.G. (1993) Total quality management and corporate culture: a case study, *Int. J. Vehicle Design*, **14**, No. 2/3, pp. 103–117.

20   Millican, A. (1995) Our commitment to quality. *Fastrack*, April, pp. 28–29.
21   Dale, B.G. and Plunkett, J.J. (Eds) (1990) *Managing Quality*. Philip Allan, London.
22   Oakland, J.S. (1993) *Total Quality Management* (2nd Edition). Butterworth-Heinmann, Oxford.
23   Mann, R. and Kehoe, D. (1994) An evaluation of the effects of quality improvement activities on business performance, *Int. J. Quality and Reliability Management*, **11**, No. 4, pp. 29–44.
24   Bicheno, J. (1994) *The Quality 50*. Picsie Books, Buckingham, UK.
25   Blackburn, R. and Rosen, B. (1993) *Total Quality and Human Resource Management: Lessons Learned from the Baldrige Awards – Winning Companies*, Vol. 7, No. 3, Academy of Management Executives.
26   Mercer, D.S. and Judkins, P.E. (1990) *Rank Xerox: A Total Quality Process, Managing Quality* (Ed. Dale, B.G. and Plunkett, J.J.). Philip Allan, London.
27   Whalley, J. (1995) *Another Step Towards Total Quality*. British Aerospace (Military Aircraft Division), Warton.
28   Cullen, J. (1995) *The Rover Quality System*. Rover Group Ltd.
29   Galt, J.D.A. and Dale, B.G. (1990) The customer–supplier relationship in the motor industry: a vehicle manufacturer's perspective, *Proc. Inst. Mech. Eng.*, **204**, pp. 179–186.
30   Lloyd, A. *et al.* (1994) A study of Nissan Motor Manufacturing (UK) – Supplier Development Team Activities, *Proc. Inst. Mech. Eng.*, **208**, pp. 63–68.
31   British Aerospace (Military Aircraft Division), (1995) *Preferred Supplier Process*, Warton.
32   Coallier, F. (1994) How ISO 9000 fits into the software world, *IEEE Software*, January, pp. 98–100.
33   Davis, C. *et al.* (1993) Current practice in software quality and the impact of certification schemes, *Software Quality J.* **2**, pp. 145–161.
34   Schulmeyer, G.G. and McManus, J.I. (1987) *Handbook of Software Quality Assurance*. Van Nostrand Reinhold, New York.
35   Card, D. (1994) Statistical process control for software?, *IEEE Software*, May, pp. 95–97.
36   Thackeray, R. and van Treeck, G. (1990) Applying Quality Function Deployment for software product development, *J. Eng. Design*, **1**, No. 4, pp. 389–410.
37   Francis, R. (1993) Quality and process management: a view from the UK computing services industry, *Software Quality J.* **2**, pp. 225–238.
38   Jones, C. (1993) *Software Productivity and Quality Today: The World-Wide Perspective*. IS Management Group.
39   *Software Industry Business Practice Survey* (1993) Price Waterhouse.
40   Burgess, A. (1994) Studies to try to quantify development trends, *IEEE Software*, January, pp. 101–105.
41   British Aerospace (Military Aircraft Division) (1995) Stakeholder analysis, *Quality CQI Training Manual*.
42   Ishikawa, K. (1976) *Guide to Quality Control*. Asian Productivity Organisation, Tokyo.
43   Asaka, T. and Oeki, K. (Eds) (1990) *Handbook of Quality Tools*. Productivity Press, Cambridge, MA.

44 Oakland, J.S. (1990) *Statistical Quality Control* (2nd Edition). Butterworth-Heinmann, Oxford.

45 Andreasen, M.M. and Hein, L. (1987) *Integrated Product Development*. IFS Publications/Springer-Verlag, London.

46 Norell, M. (1993) The use of DFA, FMEA and QFD as Tools for Concurrent Engineering in Product Development Processes, *ICED 93,* The Hague, August.

47 Miles, B.L. and Swift, K.G. (1991) Design for manufacture and assembly, Paper presented at the *IMechE Autotech 91 Congress (Seminar 33)*, Birmingham, UK.

48 Batchelor, R. (1995) (pers. comm.).

49 Shimada, J. (1992) Design for manufacture tools and methods: the Hitachi Assemblability Evaluation Method (AEM), *Proc. 24th FISITA Congress*, Institution of Mechanical Engineers, London.

50 Boothroyd, G. and Dewhurst, P. (1989) *Product Design for Assembly*. Boothroyd Dewhurst Inc., Wakefield, RI.

51 Miles, B.L. and Swift, K.G. (1992) Design for manufacture and assembly, *Proc. 24th FISITA Congress*, Institution of Mechanical Engineers, London.

52 Andreasen, M.M. (1992) Design for quality, *J. Eng. Design*, **3**, No. 1, pp. 3–5.

53 Siemens/IWF Brunswick Technical University–Quality Assurance Programme 1992–1996 (DTI Translation 0219-92). Federal Ministry of Research and Technology, Germany.

54 Swift, K.G. (1991) *Techniques in Design for Manufacture and Quality in Product Introduction–A Review of Business Practice*. University of Hull.

55 Hubka, V. (1992) Design for quality and design methodology, *J. Eng. Design*, **3**, No. 1, pp. 5–16.

56 Mørup, M. (1994) *Design for Quality*. PhD Thesis, Institute for Engineering Design, Technical University of Denmark, Lyngby.

57 Mørup, M. (1993) Design for Quality, *ICED 93*, The Hague, August.

58 Swift, K.G. and Allen, A.J. (1992) Techniques in design for quality and manufacture, *J. Eng. Design*, **3**, No. 1, pp. 81–89.

59 Swift, K.G. and Booker, J.D. (1995) *Conformability Analysis Workbook*, University of Hull.

60 Phadke, M.D. (1989) *Quality Engineering using Robust Design*. Prentice Hall, NJ.

61 Seder, L.A. (1950) Diagnosis with Diagrams – Part 1, *Industrial Quality Control*, January, pp. 11–19.

62 Batchelor, R. and Swift, K.G. (1966) Conformability Analysis in Support of Design for Quality, *Proc. Inst. Mech. Eng.*, **210**, pp. 37–47.

63 Grant, E.L. and Leavenworth, R.S. (1988) *Statistical Quality Control*. McGraw Hill, New York.

64 Buzzell, R.D. and Gale, B.D. (1987) *The PIMS Principles – Linking Strategy to Performance*. The Free Press, New York.

65 Womack, J.D., Jones, D.T. and Roos, D. (1990) *The Machine that Changed the World*. Rawson Associates, New York.

66 Schick, P. (1992) *Quality Increases Competitiveness, Quality Europe*. Carl Hanser Verlag, Munich.

67 Plunkett, J.J. and Dale, B.G. (1988) Quality costs: a critique of some economic cost of quality models, *Int. J. Prod. Res.*, **26**, No. 11, pp. 1713–1726.

68 BS 6143: Part 2. (1990) *The Guide to the Economics of Quality: Prevention, Appraisal and Failure Model*. British Standards Institution, London.

69 Plunkett, J.J. and Dale, B.G. (1991) *Quality Costing.* Chapman & Hall, London.

70 Hagan, J.T. (Ed.) (1986) *Principles of Quality Costs.* American Society of Quality Control.

71 Wood, D. and Croxall, S. (1989) Failure mode and effect analysis, Paper presented at the *Conference on Tools and Techniques in Quality Engineering,* Manchester, UK, October.

72 Ireson, W.G. and Coombs, C.F. (1988) *Handbook of Reliability Engineering and Management.* McGraw Hill, New York.

73 O'Connor, P.D.T. (1991) *Practical Reliability Engineering* (3rd Edition). Wiley, Chichester.

74 Moroney, M.J. (1990) *Facts from Figures.* Penguin, London.

75 BS 600: (1935) *The Application of Statistical Methods to Industrial Standardisation and Quality Control.* British Standards Institution, London.

76 Bird, D. and Dale, B.G. (1995) The use of statistical process control in the manufacture of high-integrity products, *Proc. Inst. Mech. Eng.*, **209**, pp. 25–31.

77 Juran, J., Gryna, F. and Bingham, R.S. (Eds) (1979) *Quality Control Handbook* (3rd Edition). McGraw Hill, New York.

78 Parnaby, J. *et al.* (1991) *Manufacturing Systems Engineering – Mini Guides.* Lucas Industries plc, Birmingham, UK.

79 Straker, D.A. (1995) *Toolbook for Quality Improvement and Problem Solving.* Prentice Hall, London.

80 Taguchi, G. (1986) *Introduction to Quality Engineering.* UNIPUB/Quality Research Resources, New York.

81 Taguchi, G. (1987) *System of Experimental Design*, Vols 1/2. Kraus, New York.

82 Bhote, K.R. (1991) *World Class Quality.* American Society of Quality Control.

83 Clausing, D.P. (1986) *Quality Function Deployment.* Mimoe Report, February.

84 Clausing, D.P. and Pugh, S. (1991) Enhanced quality function deployment, Paper presented at the *Int. Conf. on Design and Productivity*, University of Missouri-Rolla, Honolulu, Hawaii.

85 King, R. (1989) *Better Designs in Half the Time.* Goal/QPC.

86 Sullivan, L.P. (1988) Policy management through quality function deployment, *Quality Progress*, June, pp. 39–50.

87 Suzaki, K. (1987) *The New Manufacturing Challenge.* Free Press, New York.

88 Dale, B.G. (1992) Total Quality Management: What are the Research Challenges?, *Proc. 7th OMA Conf.*, Manchester, UK, June (1992), Elsevier Science Publishers.

89 Lascelles, D.M. (1990) *Research Policy.* European Foundation for Quality Management, Brussels.

Sample questions for students

The sample questions listed below provide some elemental ideas for examination questions and studies for students of engineering and business.

1 Outline what is meant by product quality and why is it so important?

2 Discuss briefly the contributions of Deming and Juran to advances in quality assurance. Include what you regard to be the strengths and weaknesses of their philosophies.

3 Discuss what is meant by the term 'total quality'. What distinguishes a total quality company from more traditional business?

4 What is meant by culture in the context of a business operation and what effect does it have on quality?

5 What influences cultural values in a business? What can be done to improve organisational culture and how should the process be started?

6 State the principles arising from the pioneering work of the quality gurus that you believe are most effective in achieving improvements in quality.

7 What parts do 'certification' and 'quality-management systems' play in the attainment of total quality culture?

8 Why is an external supplier quality strategy important in a manufacturing business and what would be the key elements needed to improve supplier quality?

9 Define a preferred supplier process and discuss the main elements involved and the expected benefits.

10 How is the quality of software defined and how can it be assured?

11 Briefly describe the QFD process and define its four phases. What are the main benefits and drawbacks associated with its application?

12 Prepare a basic QFD phase 1 matrix for the design of a foot-pump for a motor car. The matrix should include a points rating for each of the product features and functions.

13 Briefly describe the roles of QFD, FMEA, DFQ and DOE in quality improvement. Propose a programme for their application in product development showing at what stage they would be used in the cycle.

14 Discuss the Failure Mode and Effect Analysis (FMEA) methodology, highlighting the various factors that are assessed and produce a general occurrence, severity and detectability ratings table. Perform an FMEA on the prototype handlebar design shown in Figure Q14 considering only the section in View 2 and limiting the number of failure modes to four. Discuss the results of the analysis and any possible improvements to the design.

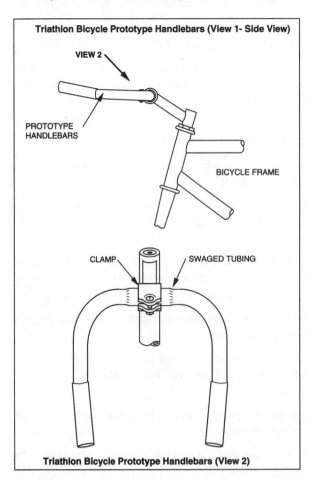

Fig. Q14 Prototype handlebar design

15 A company produces nuts by machining hexagonal bar stock. Thirty samples of four consecutively produced nuts were measured for their length when the manufacturing process was known to be running sat

isfactorily. From each sample, the average length and average range of lengths was determined, from which the following parameters were calculated:

Grand mean, $\bar{\bar{x}}$ = 9.050 mm

Mean of sample ranges, $\bar{w}$ = 0.052 mm.

Produce mean and range charts which could be used for future process control, including the control limits.

(Answers: Mean control chart – UAL = 9.075 mm, UWL = 9.089 mm, LWL = 9.025 mm, LAL = 9.011 mm. Range control chart – UAL = 0.134 mm, UWL = 0.100 mm, LWL = 0.015 mm, LAL = 0.005 mm)

16 Explain how Pareto analysis and cause-and-effect diagrams can help with the application of an SPC programme in manufacturing or technical functions.

17 You are asked to put on a training programme in SPC for an SME manufacturing engineering components for the aerospace industry. Outline the content and structure of the programme you would provide given that the training provision is to be scheduled over one year. The focus of the programme is to be on the manufacturing function.

18 What are the typical sources of out-of-tolerance variation in component manufacturing and assembly processes?

19 How is process capability quantified and how are the calculated measures of performance related to defect probability?

20 The automotive component shown in Figure Q20 is manufactured by machining. Using the SPC data in Table Q20, calculate the process capability index (C_{pk}) and state what this is likely to infer

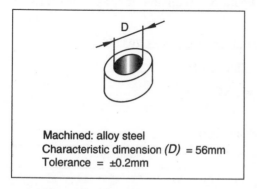

Machined: alloy steel
Characteristic dimension *(D)* = 56mm
Tolerance = ±0.2mm

Fig. Q20 Component details

Table Q20 SPC data

Sampled dimension (D)	Frequency (f)
55.78	1
55.82	2
55.88	3
55.90	4
55.94	4
55.96	8
55.98	7
56.04	11
56.06	5
56.10	3
56.16	2

in terms of defect rate in parts per million using Figure 9.15 of Chapter 9.

(Answer: C_{pk} = 0.75, failure rate = 11 000 ppm)

21 Explain why it is worth giving consideration to quality in the product design process. What design-related factors set limits on the levels of conformance that can be achieved by the application of best practice in manufacturing operations?

22 An automatic gearbox system has a set of critical tolerance characteristics, in manufacture and assembly, associated with one of its failure modes having an FMEA severity rating of 8. In production, process capability indices (C_{pk}) for the set of characteristics are listed in Table Q22. With reference to the conformability map (Figure 7.9, Chapter 7), comment on the acceptability of the shop-floor C_{pk} values and calculate the approximate total annual failure costs connected with the listed set of characteristics, based on a gearbox system selling price of £100 and a call-off rate of 100 000 per annum. What level of C_{pk} would be acceptable for the characteristics in question?

(Answer: total component failure costs = £1 793 000

total assembly failure costs = £1 921 000

acceptable C_{pk} = 1.55 − 1.8)

Table Q22 C_{pk} values for gearbox characteristics

	Component					Assembly				
Characteristic	c1	c2	c3	c4	c5	a1	a2	a3	a4	a5
C_{pk}	1	2	1.6	1.5	1.8	0.7	1.66	2.1	1.4	1.55

23 What are the main costs of quality in a manufacturing business and how can they be classified? How would you expect the quality-related costs to vary with quality awareness and improvement?

24 A major automotive manufacturer with sales of £5 billion per annum has quality costs running at approximately 22% of total revenues, and are broken down as:

- Prevention costs = 14%
- Appraisal costs = 30%
- Failure costs = 56%.

Prepare a plan for effecting a quality improvement. The plan should have clearly defined and measurable objectives and should state what financial benefits would result from your proposed course of action.

Appendix 1
Means to achieve a design review

Design Review: Why?

- To establish total understanding of the task to be done
- To establish total acceptance of task done.

Design Review: What?

Formal meeting with chairman and minutes form part of the traceability of the work. Participants need an agenda and documentation before the decisions of the meeting are binding.

- Task definition
- Who does what?
- Requirements – contractual/legal/company
- Constraints
- Achievements
- Nonconformities
- Acceptance by all involved.

Design Review: When?

Design reviews must be included in formal work plans:

- At inception, to establish understanding of requirements
- At delivery, to establish acceptability of work

and between planned life cycle stages to:

- Ensure understanding remains valid
- Ensure redirection is controlled
- Ensure problems are visible.

Design Review: Who?

- Customer(s) and supplier(s)
- Chain (knock-on effect)
- Varies according to life cycle stages
- Numbers depend on level.

Design Review: How?

- Authorisation of task
- Contract documentation (or derivation)
- Specifications
- Statement of work
- Achieved result
- Nonconformity statement
- Options examined
- Checklist used
- Procedures used.

Design Review: Output

An unambiguous definition of the work to be done
or
A statement of work completion with succeeding actions defined.

Design Review: Hierarchy

- Form a hierarchy
- Break down work to smaller components
- Build results up to a completed whole
- Form a traceable work path
- Allow visibility and control.

Design Review: Does not

- Consider costs and timescales
- Replace daily interchange between engineers
- Change departmental control of work.

These aspects must continue outside the design review and are reinforced by the formality of design reviews.

Design Review: Summary

- Is an AQAP requirement
- Is a formal procedure
- Establishes understanding of work
- Establishes acceptance of completion of work
- Requires agreement of all involved
- Forms a traceable record of technical performance versus requirement
- Forms a hierarchy of inception to completion
- Makes the whole work process visible.

Appendix 2
Quality cost categorisations – POCs and PONCs

The following lists of POCs and PONCs for the range of functions are not definitive or absolute but are given to promote discussion and debate.

Procurement	Quality
Price of Conformance	
Supplier review and approval	Training and procedures
Updating supplier specs	Quality planning
Supplier seminars	Product audit
Forecasting	System audit
Anticipating supplier problems	Design review
Vendor rating	Supplier appraisal
	Process capability studies
	Calibration
	Operator certification
	Life testing
	Test equipment review
	In-process controls
Price of Nonconformance	
Supplier scrap	Rework analysis
Supplier rework	Warranty cost analysis
Re-inspection due to rejects	Concessions analysis
Premium freight costs	Materials review board
Cost of returning goods	Review of returned material
Trips to supplier to resolve	Receiving inspection
Expediting costs	In-process inspection
Loss of supplier credit	Final inspection
Material review board	Final test

Accounts	Engineering
Price of Conformance	
Forecasting	Design specification reviews
Training and procedures	Product qualification
Ledger review	Drawing checks
Planning and budget generation	Supplier product evaluation
Supplier installation approval	Cost-of-quality budget
Special test fixtures	Capital expenditure reviews
Verifying workmanship	Order entry review
Review of test specs/FMEA	FMEA
Invoicing review	Packaging qualifications
Financial report review	Customer interface
	Safety reviews
	Defect prevention programmes
	CAD
	Training and procedures
	Development testing
	Personnel appraisal
Price of Nonconformance	
Invoicing errors	Warranty expenses
Out-of-control inventory	Engineering time on redesign
Incorrect accounting errors	Material review board
Bad debts	Design-related product reliability
Payroll errors	Failure analysis
Error in letters of credit	
Overdue accounts	

Personnel	Systems

Price of Conformance

Screening applicants	Software planning
Interviewing and selection	Software reliability prediction
Personal testing (physical and IQ)	Systems analysis
Security clearance	Documentation reviews
Introduction courses	Establishing customer requirements
Training	Operator training
Safety	Software back-up
Security	Programme testing
Attendance monitoring	Code verifying
Personnel records	Depreciation of software
Job descriptions	
Union communication	
Resource strategy	

Price of Nonconformance

Turnover rates	System failures
Absenteeism	Customer requirement re-evaluation
Employee discipline	Debugging
Grievances	Document changes
Time lost due to accidents	
Temporary labour	

Manufacturing	

Price of Conformance	Price of Nonconformance
Preventative maintenance	Scrap
Environment control	Rework
Training	Machine downtime
Procedures	Overtime excess to budget
Review of production rates	Missed schedules
Timely machine replacement	Progress chasing
SPC	Manufacturing-associated product liability
Quality circles	Obsolete and redundant stock
Manufacturing planning review	Excess overtime
Supervision	Excess absenteeism
Machine-tool alignment testing	Disciplinary measures
Tooling inspection and test	Lost time due to accidents
Process capability studies	Labour variances
Control of critical storage	
Housekeeping	
Source inspection	
Stock audits	

Index